RECENT ADVANCES IN FISH ECOLOGY, LIMNOLOGY AND ECO-CONSERVATION

VOLUME III

Editor - in - Chief
SURENDRA NATH

1994

DAYA PUBLISHING HOUSE
DELHI - 110035

ISBN 81-7035-131-6

Published by
Daya Publishing House
1123 / 74, Deva Ram Park,
Tri Nagar, Delhi-110035.
Phone: 7231826, 7220593

Laser Composed by
Delite Printers
Tri Nagar, Delhi.

Printed by
D.K. Fine Art Press
Delhi.

Bound by
Book Wrap, Delhi-110035

PRINTED IN INDIA

To
My Alma Mater
The University of Lucknow
Lucknow
India

PREFACE

The idea of bringing out Recent Advances in Fish Ecology, Limnology and Eco-conservation was mooted at Lucknow in 1966 to cope up with the latest and original researchwork in the vast fields of fish ecology, limnology and eco-conservation being carried out in various Indian Universities and research institutes. The first volume - the Golden Jubilee Commemoration volume of Professor S.M.Das, D.Sc., F.Z.S. (London), Scientist Emeritus, Lucknow - was brought out in 1988 to felicitate Professor Das on the completion of his fifty years of active research and teaching at the Universities of Lucknow, Kashmir and Kumaun.

Following upon the decision to bring out the publication annually, the second volume of the series appeared in the year 1992 and the present volume is the third in the series. It has now been decided to also incorporate in the book original research work being conducted in various foreign universities and research laboratories and the response from scientists and research workers based abroad has been quite encouraging and spontaneous in this regard.

Surendra Nath
Editor-in-Chief

Department of Zoology
The Islamia College of Science & Commerce,
SRINAGAR - 190002
(Kashmir) INDIA

CONTENTS

CONTENTS

CONTRIBUTORS

1. Dr. Tej Kumar Shrestha
Professor of Zoology,
Tribhuvan University,
Kathmandu, NEPAL

2. Dr. S.M.Das
Scientist Emeritus,
Aloka Niwas,
New Hyderabad,
Lucknow, INDIA

3. Dr. Surendra Nath
Professor & Head,
Department of Zoology,
The Islamia College of Science & Commerce,
Srinagar (Kashmir), INDIA

4. Dr. Bansi L. Koul
Head, Department of Zoology,
Govt. Gandhi Memorial Science College,
Jammu (Tawi), INDIA

5. Dr. D.K.Massey
Department of Zoology,
Christ Church College,
Kanpur University,
Kanpur, INDIA

6. Dr. A. Wanganeo
Department of Limnology,
Bhopal University,
Bhopal (M.P.), INDIA

Rec. Adv. Fish Ecol. Limn. Eco-conserv.
III :1-26

1

Status, Ecology and Behaviour of Fishes of Arun River (Nepal)

TEJ KUMAR SHRESTHA

ABSTRACT

The Arun river is a wild and scenic river of Nepal Himalaya. Some interesting features of distribution, abundance, ecology and behaviour of the rheophilic fishes of the Arun river are described. Impact of proposed upper Arun project on fish fauna and management implication are reviewed. Recommendation is made for the preservation of biodiversity and gene pools of the fishes living in the unspoiled running water of the Himalaya.

INTRODUCTION

Fishes of Himalayan waters are biologically diverse. Running waters of Nepal support more than 130 species of fishes (Shrestha 1990). The scientific study of their kind, occurrence and distribution in different drainage systems is not a task of top priority to formulate their conservation and management policies. Taking this fact into consideration an ecological study of the fishes occurring in upper reaches of Arun river was attempted to advance our knowledge about the habit and behaviour of fishes living in the running waters of the Himalaya.

THE STUDY AREA

The Arun river lies in between Nepal and China border and is one of the unique ice cold rivers of Nepal known to naturalists and explorers. The river has a spectacular gorge in Himalaya lying between Nepal-China border, which is famous all over the world. The Arun river originates in Tibet, where it flows approximately 129 kms before it joins its feeder streams Sunkoshi and Tamur at Tribenighat of Nepal. Along its course in Nepal it receives countless feeder streams of different sizes, creating excellent spawning and nursery areas for resident and migratory fishes.

METHODOLOGY

For the study of riverine habitat both direct and remote sensing methods were used. Direct sensing of the mahseer habitat was made with unaided eye. Remote sensing of the habitat was done by using polarized eye glasses. Observations were aided by binoculars (7 × 35) and notes were recorded in a portable tape recorder for future analysis. Among the methods used were : (a) information sampling, (b) direct observations, (c) photography, and (d) netting at suitable places.

The shape and size of spawning ground and physical characteristics of water was studied by using geological and hydrological maps. Possible spawning habitats were identified and

Fig.1 : Arun Gorge Near Gola Village at Dharmatala. This barrier precludes heavy bodied fish from entering.

collection spots for the study were fixed. The spawning habitat condition, e.g. type, cover, shelter, obstructions and water level were recorded.

Various kinds of locally prepared fishing nets were used for fish sampling. For example, cast nets were used for sampling different sizes of immature fishes. Scoop nets and dip nets were used for capturing larvae and fingerlings. To collect fishes from deep pools fish-snaring hook lines were used. To collect fries of fishes, under water torch light, petromax (300 to 500 candle power) were used to attract them at night near the bank of the river and this technique was found to be very effective.

Studies on the water quality parameters such as temperature, turbidity, dissolved oxygen, H-ion concentration, carbonates, bicarbonates and alkalinity of surface waters were conducted. Dissolved oxygen was determined by modified Winkler method. The current velocity and pH was measured by current meter and pH meter. Turbidity was determined by Secchi disc and water temperature by a 50°C thermometer.

Live fishes were collected and dissected. Samples of ovaries and testes were drawn and their maturity stages were examined under a binocular microscope. The ova-diameter was measured by means of a micrometer.

The activity of fishes was sensed by sight or by watching fish from water level above or from beneath. Also fish watching was done by polarized eye glasses or binoculars. Every visual observation was supplemented or replaced by photography. Fish photographs were made by under water camera (Nikons - IV) model which gave better results.

FIELD STUDY

(a) Specimen Collection

The time allocated in the task assignment for the field investigation was 40 days between June-August. The fish species collected were measured to the nearest millimeter for total length and standard length and weighed in grams. Each species of fish was dissected to take out gonads, and in most of the cases, the weights were also taken. Some specimens of fish were preserved in 10 per cent formalin for taxonomic identification purpose.

Aquatic vegetation such as algae (both floating and attached) were collected and preserved in formalin. Varied aquatic crustaceans and insects were collected and preserved in 70 per cent alcohol for identification. Stomach contents of some fishes were taken out and preserved in dilute formalin for microscopic analysis.

(b) Laboratory Analysis

The specimens of fish collected from the different sampling areas were identified using standard methods of taxonomy (Jayaram 1981). The specimens of algae and invertebrates were also identified nearest to genus and stomach contents were analyzed, most of which were fully digested, but some fragments of algae and invertebrates which were intact were identified. Sand and silt collected from the rivers were measured through micrometer attached to the microscope.

(c) Fish Migration and Population Dynamics

The spawning migration was observed by watching fish from water level above or from beneath. Also fish watching was done using polarized eye glasses or binoculars. Every visual observation was supplemented or replaced by under-water photography.

1. The spawning migrants were observed close to the surface of the pool and their exact position, orientation and action were recorded at each time. The same pool was observed several times to acquire data of the change in brood fish stock for a week.
2. The degree of health and liveliness of brood fish or migrants caught in net traps was noted down and their fins were clipped for marking purpose.
3. Migratory activity and mobility of migrants was photographed with under-water camera, using various filters.
4. Four feeder streams or creeks joining the Arun river were diverted, dewatered and local ichthyotoxic herbs were used for the collection of voucher specimens.
5. For recording fishing activities, nocturnal aquatic animals such as river otter strong electric tubelight and torch light was used.

LITERATURE REVIEW

Fish fauna of Arun river is biologically diverse. First authentic information about the kind of occurrence and distribution of fish in Suklaphanta Wildlife Reserve was given by Shrestha (1976). Later on Shrestha, Shrestha and Pradhan (1979) studied aquatic ecology and potential fishery development in the downstream section of the Terai. Information concerning ecology of the fishes of Kathmandu valley is now available from Shrestha (1979). Bhattarai (1977) studied rural water supply development in Nepal. Sharma (1978) studied the quality of the drinking water of Kathmandu. Shrestha (1980) studied biological indicators of pollution of the river Bagmati. There is little information on the fish fauna occurring on the upper reaches of the Arun river. Therefore, according to the term of reference by an intensive sampling, collection of fishes occurring in the upper Arun section was conducted to provide access to fish fauna and their habitat.

Fig. 2 : Arun River near Sibrung village. The area is rich in snow trout.

A considerable amount of literature exists on dams and their impact on cold water fisheries in other parts of the world, (Low-Mcconnel 1976, Lagler 1969 and 1976, Letriz and Lewis 1976, Ruggles 1980). The scientific management of the cold-water fish living in mountain stream has just begun. Also, information on cold-water fishery of cold-water indigenous fish living in equivalent condition of torrential water of the Arun river is meagre. In Nepal, much aquaculture work is concentrated on exotic carps and warm water carps that can adapt to temperature regime of the Terai region. In recent past the present author (Shrestha 1978, 1979, 1986) has undertaken some detailed studies on reproduction and breeding potential of some cold water species occurring in Panauti and Trisuli rivers. His study relates in general to the environment, specially to the region within which Arun is located.

HYDROLOGY

The Arun is the main river of the Sapta Kosi system located between latitudes 26^0, 42' to 80^0 57' E. It originates from a glacier on the north slope of Mt. Xixabangma Feng (8012m) and is called the Pum Qu within Tibet. It flows eastward almost parallel to the Himalayan range in the upper reaches for a distance of about 280 km and then takes a sharp turn to the southwest at the junction with a large tributary (Yeyuzangbu river), forming a big bend. The Arun then flows southward crossing the Himalayan range into Nepal. It continues to flow south and joins the Sun Kosi and Tamur rivers at Tribeni. The total length of the Arun is about 3,400 km, of which more than 95 per cent lies in Tibet.

Within Tibet, the river can be divided into three distinct reaches (Interim Report, May

Fig.3 : Arun River just below Adima Village. This Zone is rich in stone carps, catfishes and Loaches.

1991). In the upper 150 km, the river flows with an average gradient of about 1/130, with terraces scattered along both banks. Within the next 130 km reach, the river-channel slope becomes milder with an average gradient of about 1/630. The river is broad with a wide valley and flood plain and the channel is braided and meandering. In the lower 100 km reach of the river, before it crosses the Himalayan range to Nepal, the channel becomes steep with an average gradient of about 1/50 to 1/30, forming a deep V-shaped gorge. The average slope along the reach between the Nepalese-Tibetan border and the damsite is about 1/30.

The proposed Upper Arun damsite is located in the narrow gorge approximately 350 metres upstream of the confluence of Chepuwa Khola waterfall and Arun river and about 14 km downstream of the Tibetan border. The powerhouse site is located some 15 km downstream of the confluence of the Barun and Arun rivers. The drainage areas evaluated from hydrological investigations are given as table 1.

Table - 1

Arun River Basin	Drainage Area (km^2)
Upstream of border (Tibet)	25,300
From border to Chepuwa gauging station (about 500 m upstream of damsite)	400
From border to powerhouse site	970
From border to Uwagaon	1,320
From border to Tumlingtar	2,480

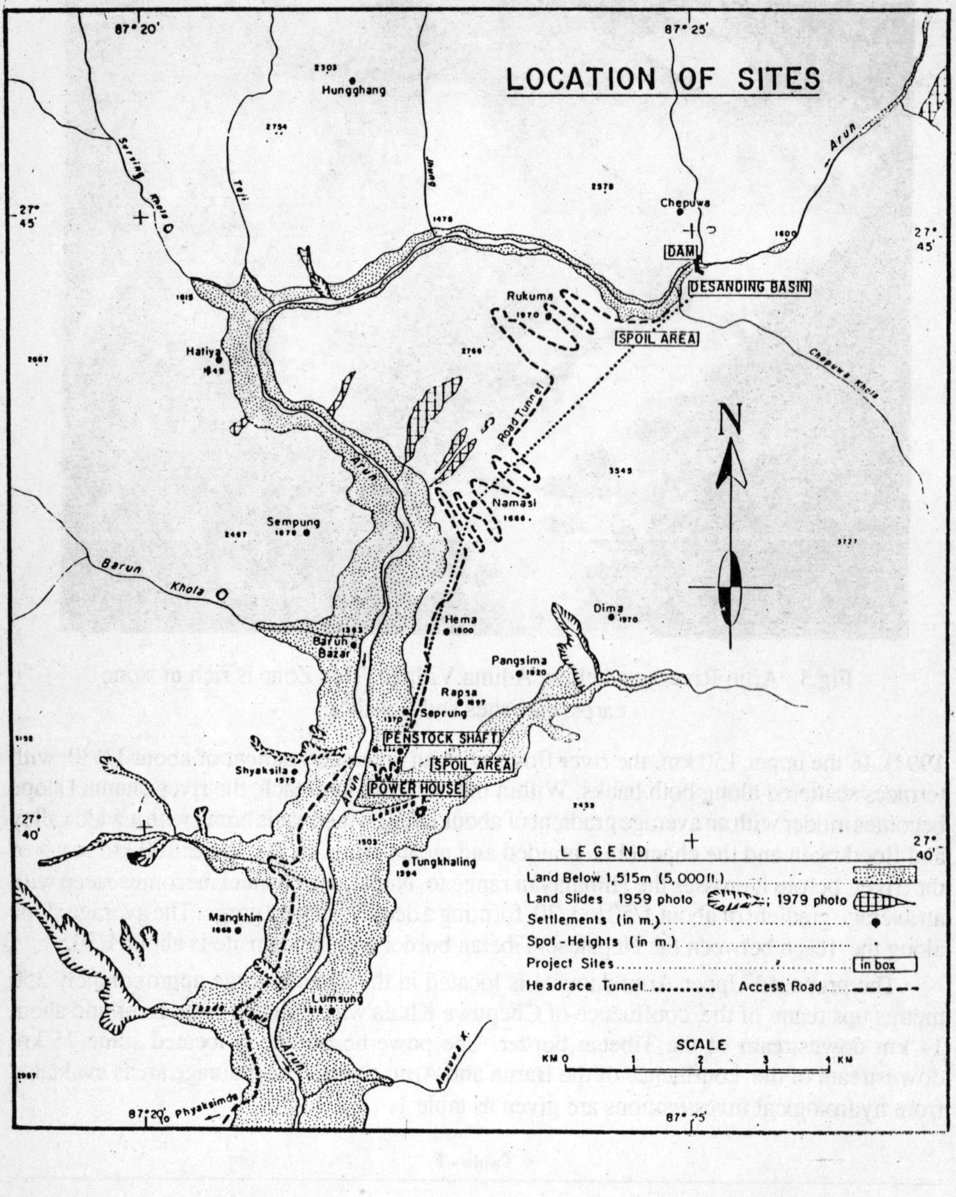
LOCATION OF SITES
87° 20'
87° 25'
27° 45'
27° 40'
Hungghang
Chepuwa
DAM
DESANDING BASIN
Rukuma
SPOIL AREA
Chepuwa Khola
Arun
Sursing Khola
Tolji
Hatiya
Road Tunnel
Namasi
Sempung
Barun Khola
Hema
Dima
Baruh Bazar
Pangsima
Rapsa
Seprung
PENSTOCK SHAFT
SPOIL AREA
POWER HOUSE
Shyaksila
Tungkhaling
Mangkhim
Lumsung
Amsuwa K.
To Phyaksinda
N
LEGEND
Land Below 1,515m (5,000 ft.)
Land Slides 1959 photo ; 1979 photo
Settlements (in m.)
Spot-Heights (in m.)
Project Sites in box
Headrace Tunnel
Access Road
SCALE
KM 0 1 2 3 4 KM

Map I

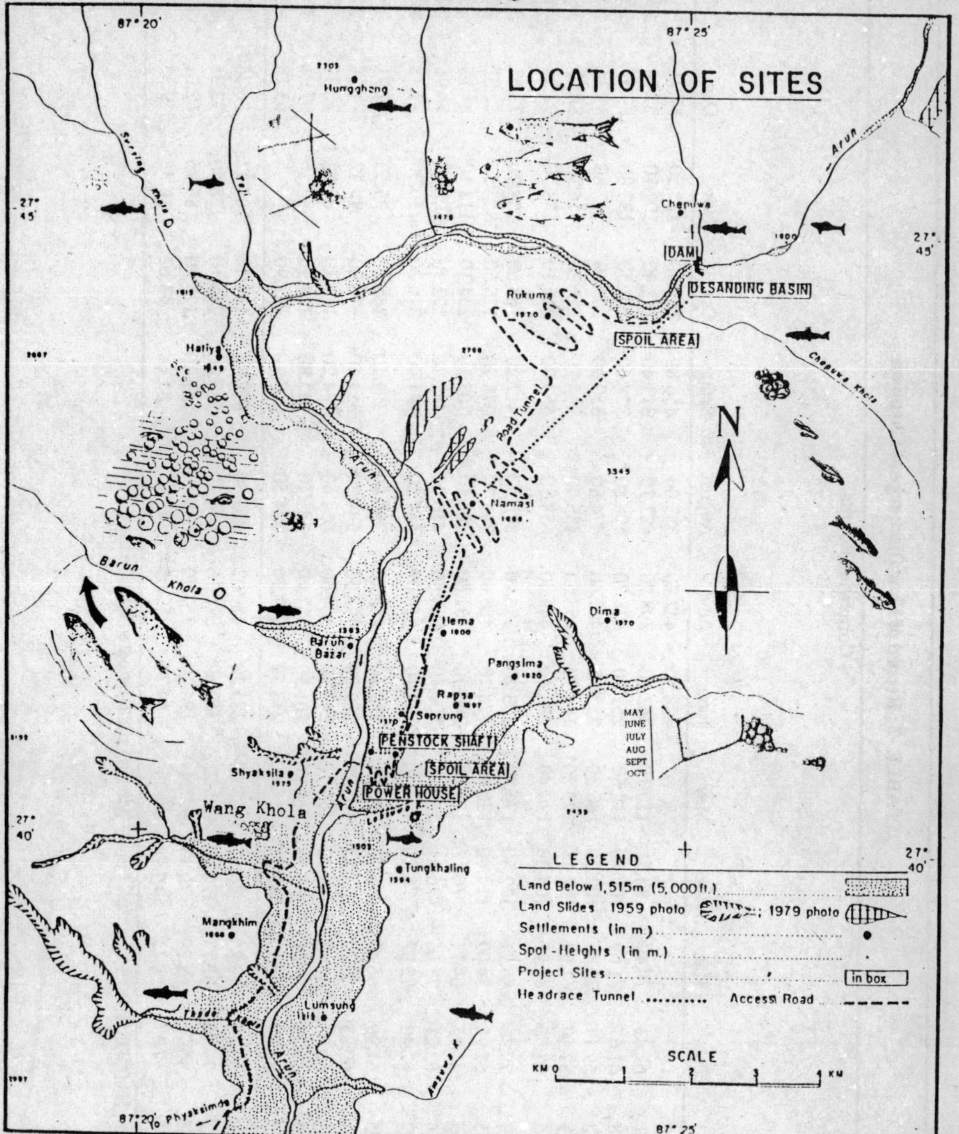

Tributaries Used by Migrating Fish

Spawning Grounds

Map II

TABLE -2 : Record of Flow Discharge at Chepuwa

Correlation Results

Discharges (m3/s)

Year	Jan.	Feb.	Mar.	Apr.	May	June	July	August	Sept.	Oct.	Nov.	Dec.	Yearly Mean
1975	-	-	-	-	-	288.30	442.20	385.00	456.80	257.50	127.50	87.91	-
1976	69.41	67.43	70.49	80.51	129.10	362.60	398.70	422.70	313.40	172.40	109.10	77.18	190.07
1977	61.48	61.67	71.69	100.20	132.50	253.70	428.70	508.80	328.60	232.30	133.80	89.41	201.24
1978	70.72	69.17	78.40	102.30	180.30	355.10	423.40	474.80	350.90	203.60	117.80	81.36	209.78
1979	60.53	57.19	62.89	97.78	129.30	194.10	458.50	435.40	344.70	232.20	114.30	84.67	189.49
1980	66.35	67.56	81.91	104.30	130.40	282.10	449.40	638.30	553.00	265.10	138.50	92.38	240.11
1981	70.23	79.36	88.77	122.80	179.30	333.20	526.10	573.50	417.20	168.60	104.20	73.63	229.02
1982	61.75	62.98	84.09	114.00	136.00	328.20	486.60	391.70	278.70	125.50	85.95	62.01	185.52
1983	49.65	46.75	64.14	70.17	129.50	246.30	481.60	408.50	505.10	244.30	113.10	81.13	204.20
1984	64.84	62.21	75.66	75.11	151.80	307.40	524.50	416.20	441.70	192.30	102.10	74.95	208.23
1985	59.82	62.49	73.99	78.00	105.80	281.10	515.00	451.50	441.40	216.00	110.30	82.78	207.37
1986	64.75	61.99	66.42	83.77	92.24	289.60	482.20	396.80	309.10	166.20	95.53	67.91	182.13
1987	58.41	60.33	68.19	84.23	102.20	289.00	495.50	544.30	359.50	176.00	99.12	71.28	210.65
1988	60.81	60.08	73.45	88.63	117.10	243.70	504.60	654.40	363.20	146.20	88.88	70.37	207.20
1989	63.12	69.23	83.48	85.55	164.10	261.50	357.70	345.40	312.80	154.90	86.16	67.90	171.58
1990	58.17	58.79	64.15	80.65	115.80	248.50	369.70	374.50	257.10	141.90	-	63.13	-
Mean	62.70	63.15	73.85	90.53	133.03	285.28	459.02	463.86	377.10	193.44	108.42	76.78	199.77
Max	70.72	79.36	88.77	122.80	180.30	362.60	526.10	654.40	553.00	265.10	138.50	92.38	240.11
Min	49.65	46.75	62.89	70.17	92.24	194.10	357.70	345.40	257.10	125.50	85.95	62.01	171.58
STDV	5.48	7.17	8.11	14.98	26.26	44.74	52.66	94.27	83.68	43.62	16.38	9.33	18.05

Source of Raw Data : HMGDMH 90 Missing data indicated by a Dash

Stream Flow

Three stream-flow gauging stations are under operation on the Arun river. The Chepuwa gauging station is located about 500m upstream of the Upper Arun damsite. It has been operated since January, 1989. The Uwagaon gauging station, situated some 25 km. downstream of the dam site started recording in 1985. Gauging height readings are available for 1972 to 1985, however, without corresponding flow measurements. The only gauging station which provides continuous stream flow records since 1975 is at Tumlingtar, about 70 km downstream of the damsite.

The effective drainage area was assumed to consist of about 400km^2 in the Nepalese portion upstream of the dam site of about 4000 km^2 in the wet season and 10,000 km^2 in the dry season in Tibet. The relatively large rainfall in the downstream part of Arun river basin on the Tibetan side that effectively contributes to the run off in both seasons was taken into account. Thus, the average annual discharge from the synthesized flow series was 280m3/sec and the daily discharge exceeded 95 per cent; 100 per cent of the time was about 90m3/sec and 78m3/sec respectively at the dam site.

Water Velocity

Approximately 80 per cent of the annual run-off is caused by monsoon rainfall brought about by orographic uplift of the moist air on the windward slopes of the Himalaya. The remaining 20 per cent of run off is mainly sustained by snow-melt and ground water recession in Tibet. The wet season lasts approximately from May to October and the dry season from November to April.

With only two years of recorded daily flow data at Chepuwa being available, the data series had to be extended by using flow records from Uwagaon and Tumlingtar station. The series of mean daily stream flow was as follows :

Chepuwa	(1975-1990) : 200m3/sec
Uwagaon	(1975-1989) : 344m3/sec
Tumlingtar	(1975-1988) : 417m3/sec

Monthly discharges increase considerably at all stations during the monsoon season, with greatest discharge between July and August. The increase, however, varies from station to station and reflects the varying meteorological conditions of the Nepalese drainage basin.

The intermediate catchment between Chepuwa and Uwagaon belongs to the region around Num which exhibits the highest monsoon precipitation of the Arun Basin. With increasing distance to north or south of Num, the monsoon precipitation decreases and leads to smoother annual hydrograph, in particular at Chepuwa. For the years 1989 and 1990, the hydrographs of Chepuwa were compared with those of Uwagaon. Furthermore, these exhibits show the observed maximum daily flow within each month; and these flows indicate the same tendency as obtained with the mean monthly flows, similar flows during the dry season, and significantly higher flows at Uwagaon than at Chepuwa during the monsoon season.

The highest level of calcium carbonate was found in the water sampled from Chepuwa stream, Yakpe spring, Sebrung stream and Namasi spring (117 to 119 mg/L). The WHO standard level of hardness for potable water is 100 to 150 p.p.m. CaCo2. Thus the highest level of hardness of water found was also up to WHO standards.

TABLE - 3 : Physico-Chemical Parameters of Arun Down Stream and Upstream Water *

Sampling Site	Temp. °C	pH	Turbidity Unit	Potassium Mg/L	Sodium Mg/L	Ortho-Phosphate Mg/L	Calcium Mg/L	Calcium Carborate Cacoz Mg/L	Magnesium Mg/L	Conductivity Mg/L	Sedimentation Total Solid Mg/L	Total Dissolved Solid Mg/L	Total Suspended solid Mg/L	Salinity Mg/L
1. Arun river (Tumlingtar)	12.0°C	7.0	6	1.3	1.1	0.3	16.6	74.1	7.9	137	112	68	44	0.3
2. Arum river (Power house site)	6.5°C	7.0	4	1.4	1.2	0.2	27.1	101.4	8.2	159	180	156	24	0.5
3. Arun river (Chepuva above dam site)	3.0°C	7.0	6	1.4	1.2	0.5	28.0	71.2	9.0	214	135	76	59	0.3
4. Arun river (Chepu dam site)	3.0°C	7.0	4	1.3	1.2	0.5	28.4	72.5	9.0	216	168	139	29	0.3
5. Arun river (Chepuwa below dam site)	3.0°C	7.0	3	1.4	1.2	0.5	28.4	71.6	9.0	214	129	102	27	0.4
6. Arun river (Mixing point with)	5.0°C	7.5	4	1.2	1.2	0.7	29.0	71.2	8.5	214	131	98	33	0.3

* *Source* : New Era Report

Table 4 : Results of Biochemical and Bacterial Water Tests *

Sampling Site	Dissolved Oxygen Mg/L	Biochemical Oxygen Demand Mg/L	*Bacterial Count* Total Count Colony /100 ML	Fecal coliform Fecal /100 ML
1. Arun river (Tumligtar)	16.0	0.116	5	2
2. Arun river (Power house site)	14.0	0.2	0	0
3. Arun river (Chepuwa above dam site)	13.0	3.1	0	0
4. Arun river (Dam site)	22.7	0.36	0	0
5. Arun river (Below dam site)	13.0	0.11	2.0	0
6. Arun river (Mixing point with)	21.0	0.15	5.0	1

* *Source* : New Era Report.

Low dissolved oxygen content is a sign of organic pollution. The WHO standard for dissolved oxygen is 10 mg/L with a tolerance level of not less than 6 mg/L. The dissolved oxygen level in all the tested water was very high. It ranged from 23.2 to 9 mg/L. This high content of D.O. also suggested good water quality. However, the specimens were stored for a long time.

The biochemical oxygen demand is defined as "The weight of oxygen absorbed by a given sample of water in five days at 20 degree C". The biochemical oxygen demand (B.O.D) indicates the degree of biological pollution. The B.O.D is the amount of oxygen required by bacteria while stabilizing decomposable organic matter under aerobic conditions. Water effluent should not have more than 30 mg/L if discharged into the biologically active streams. The B.O.D. of samples from Arun and Barun river waters indicate absence of organic pollution.

Chemical Tests

Chemical substances above threshold levels are harmful to living organisms. Serious pollution occurs when the water is stagnant and the microbes scavenge the available oxygen and sooner or later the anaerobic organism starts to grow and produce hydrogen sulphide. The permitted level of total coliform bacteria in drinking water as per WHO guidelines is 3 colonies/100 ml.

Waters from the Arun river, Barun river and downstream and upstream of the project site had conductivity varying from 216 to 137 mhos/cm at 25 degree C. Conductivity is a numerical expression of the ability of an aqueous solution to carry an electric current. High conductivity indicates high presence of ionic substances. The pH of water from different sites on the Arun river was almost neutral. The magnesium level was 7.9 to 9 mg/L. The amount of calcium varied from 16.6 to 29.0 mg/L, orthophosphate from 0.1 to 0.7 mg/L and calcium carbonate 101 to 71.2 mg/L. The hardness of water is due to the presence of bicarbonates of Ca^{++} and Mg^{++}. The turbidity of Arun river varied between 6 to 3 units. The total solid of Arun water is between 180 and 112 mg/L. The amount of sodium varied between 1.2 to 1.1

mg/L and potassium 1.4 to 1.2 mg/L. Salinity was also very low. All the above chemical parameters were found to be within WHO standards for potable water. The results indicate that Barun river water has a lower level of magnesium and lower hardness of water than that of Arun river water. Similarly, sodium, potassium and salinity levels of the Barun river were lower than those of Arun river water samples.

The pH of water at different areas of the Arun project site varied between 6 to 7.5, which is within the tolerance level of WHO standards for potable water. Similarly, the turbidity of spring waters was up to potable water standards. The WHO drinking water standard for turbidity is 5 units. The orthophosphate level of spring and stream water varied between 0.7 to 0.2 mg/L, magnesium 9 to 5.1 mg/L, and calcium 33 to 33 mg/L. These chemicals were found below the drinking water standard. The total hardness of water in terms of calcium carbonate varied between 21 to 119 mg/L.

Fish and Aquatic Life

The fish of Himalayan waters are biologically diverse (Hora: 1937; Menon: 1974). More than 130 species occur in the snow-fed rivers and lakes of the Nepalese highlands (Shrestha 1990). Of the major rivers, Kosi has 108, Narayani 102, Karnali 74 and Manakali 69 species (Shrestha 1990). Khadka (1989) investigated fish in the Arun river and its tributaries and reported 18 species. In the Sunkoshi and Trisuli rivers, dams and barrages have been constructed for hydro-electric power generation. These dams lack fish-ladders and obstruct or prevent upstream and downstream movement of fish. Rare migratory fish species like the mahseer and local 'snow trout' are also affected by changes in water velocity and fluctuating

Fig.4 : Baruni River, a feeder stream of Arun, supports high diversity of fish.

levels that sometimes expose their spawning beds. The shallow tailwater areas below the dams where fish congregate are exploited by fishermen, poachers and predatory animals (Shrestha 1990). Rivers in Nepal provide habitat for fish ranging from the torrential snow trout to the culturable carps. A survey of fish fauna of the Arun river near the dam site revealed 25 species belonging to seven families. Five species are of some commercial importance. Many fish species of Indo-Chinese affinity could be expected from the Arun river, but were not collected and identified due to the time of year when this study was undertaken.

Fish Species in Arun River and Tributaries

Both migratory and resident fish species were caught by using cast net, gill-net, dip-net and fish spear and hooklines. Long handled dip-net and fish spears were most productive for sampling fish in the main Arun river. A variety of locally made gradient fish traps with strong non-return value were used to trap local fish with migratory tendencies. Such traps were suitable at places where free access for netting was poor. To understand the pattern of fish distribution, the Arun river tributaries near confluence sites were diverted and fishes living in the main river and tributary were sampled.

The sampling result showed that the upper reaches of the Arun above and below the Chepuwa dam site were dominated by mountain trout while other fishes such as stone carp, cat fish and loaches occurred in subdominant level. The sampling result of the Chepuwa khola near the dam site revealed 74 per cent snow trout, 12 per cent rock carp, 10 per cent cat fish and 4 per cent stone loaches. As large deep-bodied copper-and golden-mahseer cannot swim past the rapids, they were absent in this section.

Another sampling near the Arun and Barun confluence also showed snow-trout as dominant species and other species ocurring in subdominant level. Catch in the Barun Khola was dominated by 70 per cent snow trout, followed by 15 per cent rock carp, 10 per cent cat fish and 5 per cent stone loach.

Sampling in Leksuwa-Arun confluence also showed predominance of snow trout and other fishes occurred in subdominant level. Presence of Katla was detected in the Leksuwa Khola. This gave direct evidence that Leksuwa is a spawning ground for Katla. The catch composition in Leksuwa was snow trout 78 per cent, rock-carp 10 per cent, cat fish 8 per cent, loach 2 per cent and katla 2 per cent.

A section of the Arun river below Pathibhara and Uwagaon was very productive due to gentle gradient, suitable water temperature, presence of sheltered pools and backwaters. Catch in this section was dominated by snow-trout and katla occurred in subdominant level. The catch composition in the Arun near Pathibhara section was snow trout 63 per cent, katla 25 per cent, rock carp 6 per cent, cat fish 3 per cent, Jalkapoor 2.0 per cent, fresh water eel 0.5 per cent and stone loach 0.5 per cent. The presence of Jalkapor and fresh water eel is occasional and represented in incidental catch only.

The major tributaries of the Arun in the upper section where spawning occurs are the Jhoda Khola, Wang Khola, Leksuwa Khola and Barun river, located far below the dam site. For the spawning activities of these fishes separate study is under way.

The Arun river harbors mountain snow trout (*Schizothorax plagiostomus*) in the upper reaches, Copper mahseer (*Acrossocheilus hexagonolepis*) in the middle reaches and a variety of indigenous carps in the lower section. The upper reaches near the Arun and Barun confluence between Hatiya and Gola villages have most of the hill stream fish and river carp species. At least 24 species have been recorded from upper reaches of the Arun river, of which 23 were collected during sampling in December 1990.

The principal fish of the Upper Arun are :

1. *Carps* : Snow trout (*Schizothorax richardsonii, S progastus*), Copper Mahseer (*Acrossocheilus hexagonolepis*), River rohu (*L.dero*), Minor carp (*Puntius sarana*), River barb (*Barbus chilinoides*) and Stone roller (*Crossocheilus latius*).
2. *Stone Carp* : Stone catfish (*Psilorhynchus pseudecheneis*) and rock carp (*Balitora brucei*).
3. *Catfish*: Torrent catfish (*Glyptothorax pectinopterus, G. telchitta*), Catfish (*G.trilineatus*) and *Pseudecheinus sulcatus*.
4. *Loach Species* : Stone loach (*Noemacheilus rupicola, N. beavani*), Rock sucker head (*Garra gotyla , G. annandalei*).
5. *Shad* : Baikha *(Clupisoma garua)*.
6. *Eel* : Fresh water eel (*Anguilla bengalensis*)

The following account briefly describes sections of the Arun river and its' tributaries in terms of habitat suitability.

Arun-Chepuwa Khola Confluence : The Chepuwa Khola is not congenial for fish. The course of the river is steep with a high gradient. It contains boulders, rocks and pebbles of all sizes. Fish habitat is poor and spotted snow trout is a dominant species.

Arun-Barun Confluence : The Barun Khola joins the Arun river below Hatiya village. It is larger than Chepuwa stream with bigger boulders and a higher gradient. The analysis of bottom substrate revealed that pebbles of 2.5 cm - 5.0 cm size constituted 70 per cent of the total sample. Among sand, particle sizes of 150 cm - 700 mm were more dominant. Water temperature in December at mid-day was 15°C with a pH of 7.0. Surface velocity in the main river at the confluence was 2.7-3.5 m/Sec and surface velocity in a tributary pool before the confluence was 1.4-1.9 m/Sec. *Schima wallichi* is the dominant vegetation but *Terminalia tomentosa* and *Arundanaria falcata* are also common along the banks. Filamentous algae, *Spirogyra* and *Ulothrix*, are common fresh-water algae. This section is a major spawning ground for snow trout and catfish and provides very good habitat for longnosed snow trout.

Arun-Leksuwa Confluence : Leksuwa Khola enters the Arun River below Sebrung. It has a low gradient, contains few boulders and has wide stream bed. The bottom sediments at the confluence contain nearly 70 per cent pebble particles (size range of 0-8 cm to 10 cm) and finer sands of 150 um - 1.5 um size. In December, water pH was 7.2. Water velocity recorded at the confluence was 1.0-1.7 m/Sec in rapids and at the surface of river pools it was 0.5-0.8 m/Sec. *Lindera pulcherrima* and *Acer caudatum*, are dominant vegetation along the banks, with algae like *Spirogyra Ulothrix, and Pleurococcus* as dominant. Among diatoms, *Navicula*, and *Cyambella* are well represented. This section is identified as best fish habitat in the Upper Arun.

Arun-Wang Confluence : Wang Khola receives a number of creeks and merges into the Arun near the Gola check-post. This stream is wider, contains large boulders and has a less steep gradient. Bottom sediments at the confluence are characterized by 60 per cent pebbles of the size range of 0.6-0.8 cm. and fine sand (250 um - 750 um). In December water pH was 7.6. Water velocity of the main river was 1.2-2.8 m/Sec, and in the pools 0.4-0.7 m/Sec. *Alnus nepalensis* and *Choreospondias axillaris* are the dominant riparian trees. Filamentous algae are abundant.

Distribution Patterns of Fish at Project Sites

Dam Sites

The condition of fish habitat is poor near the dam site. The river gradient and current velocity greatly limits fish species occurrence. Fishing with cast nets revealed only snow trout. Trapping revealed stone catfish (*Exostoma blythil*) and stone loach (*Noemacheilus rupicolsa*). Fishing with snaring loop-line and hooks here was very effective, as was the use of a gradient fish trap.

***Powerhouse site* (Arun-Barun confluence to Leksuwa Khola confluence) :**

This river section is a relatively good habitat for snow trout. Pools, rapids and backwaters were sampled and snow trout were found to be abundant. Besides snow trout, stone carps, rock carps and stone loaches were trapped in spindle traps. A few juvenile specimens of the torrent catfish (*Glyptothorax pectinopterus*) were caught. Netting in Sebrung Khola with fine meshed net (0.2 mm) yielded stone loach (*Noemacheilus*) and rock fish (*Garra gotyla*).

***Downstream of the Upper Arun Power House Site* (Gola to Pathibhara):**

This section provides good habitat for fish. The major tributaries are Wang Khola, Thado Khola and Kami Khola. The high density of aquatic insects provides food for fish. The indicator fish species is Katla (*Acrossocheilus hexagonolepis*) but its density is low as compared to that downstream.

***Downstream Section* (Pathibhara to Phyaksinda) :**

This section is hydrographically interesting. Numerous brooks and creeks join the Arun river to enrich fish diversity and the food chain. Many pools, rapids, backwaters and gravel-bed nurseries are present in the Arun river and its feeder streams. This section is dominated by Katla. *Sahar* fingerlings are found here but it visits only during the spawning season. Kami Khola and Panch Pokhari Khola also support a high density of snow trout, copper mahseer (Katla), rock carp and stone carp.

The tributaries of the upper reaches support snow trout or Asla (*Schizothorax plagiostomus*), and middle reaches support Katla (*Acrossocheilus hexagonolepis*). Many hill stream fishes are known to descend downstream and Tumlingtar is perhaps the lower limit of the species. Occasionally during floods, these hill stream fish may be washed down to Saptakoshi river near Barahachettra.

MIGRATORY SPECIES AND MIGRATION CYCLE

Copper mahseer

Copper mahseer were captured from few downstream sites during the summer sampling period. They are migratory species, and examination of the population size distribution of fish collected indicates that mainly large fish are present in July and August and smaller fish throughout the year. Fish egg and fry collection data and information from the spawning beds of the Arun river in September indicate the life cycle of copper mahseer of Arun river as follows :

- Mature copper mahseer with ripe and running eggs and spermary migrate upstream in late June and August as discharge from monsoon flood begins to increase. These adults enter tributary streams during early June.
- Spawning copper mahseer occur during September-October in gravel bottomed creek and streams. Adults remain in spawning ground or assembly for a short period and migrate to downstream pool area of main Arun stream and repeat several times until all eggs present in the ovary are deposited in the spawning beds. August collection yielded only large and gravid fish with mature ovary and spermary.
- Eggs hatch between the last week of September and October in nursery beds located in the banks of Arun as well as tributaries and fry and fingerlings remain until the following year. After having grown to 8-12 cm size they migrate downstream in November, usually a few weeks after the adults have migrated downstream.

Snow trout

Snow trout is a small scale migrant, well distributed in upstream and downstream of Chepuwa dam, with part of population extending to Sava Khola section of the Tumlingtar. After the creation of the dam small fraction of the latter population will become separated from upstream population but are expected to establish themselves since their spawning and nursery habitat in tributaries. Flow from tributaries during spawning season make up about 5 per cent of total Arun flow at dam site and downstream area; this should be insufficient to allow population to establish new habitat between dam site and downstream power house areas near Sibrung. Upstream population of snow trout will not be significantly affected by dam and water diversion , but the downstream population will be affected; barrier effect of dam and fish hatchery will be required to sustain population of snow trout.

Other migrants

Downstream sites near the Uwa Goan tended to harbor a wide variety of long range species such as freshwater eel (*Anguilla bengalensis*), Jalkapor (*Clupisoma garua*) and Thed (*Labeo angra*) etc. These species appear to primarily inhabit downstream sites and also make use of upstream site at dam during monsoon. The details of the spawning biology, life history and migratory cycle of these species are not clearly known.

SPAWNING HABITAT

Fish spawning habitat of Arun is less stable and unproductive due to cold water temperature and high water current. The Arun has steeper gradient upstream in Chepuwa area and more gentle gradient between Leksuwa and Wang Khola confluence site. This difference in stream morphology has implications in spawning habitat. The higher gradient upstream causes high flow of velocities, few pool areas, and make gravel-bed stream suitable for snow trout. Downstream from project area, near Uwa Gaun, the river morphology is quite different from upstream sections. Here the lower Arun has frequent sand bars, and wider flood plain, lower gradient and water velocity, and frequent pools and riffles suitable for copper mahseer and eels species. The water quality parameter of the Arun is given in Appendix III. The Project areas near Chepuwa were visited in highwater phase of September. Fish samplings were carried out in upstream and downstream habitats. Field studies of fish habitat and spawning beds were conducted in the first week of September.

Table 5 provides a species list of resident and migratory fish and the common name of fish species captured from the project area during August 1990 and September 1991.

Spawning Ground of Snow trout

In Arun river main spawning ground of snow trout was located about 10km downstream near Thudam Khola, Chepu Khola, Honkong Khola, Teji Khola, Sirsung Khola and Leksuwa Khola in September 1991. The spawning bed was composed of fine pebbles (600 to 800 mm) and gravel 300 to 500mm in size. The depth was 1-3m with deepest part 3m. The eggs and embryos were seen in shallow river bank which varies from 2 to 5.5m in depth, and it is here, that snow trout were observed in the study period of August. The water velocity ranged from 2.5 to 3.5m/Sec; temperature 22°C, pH 7.0 and dissolved oxygen 11ppm appear to be favourable for spawning. Snow trouts were seen moving to deeper water near Leksuwa confluence near Sibrimg. Such upriver movement began in early June, when water started becoming turbid due to influx of rain.

The developing eggs and larvae of snow trout were seen in shallow, semi-stagnant nursery bed along river bank interspersed with gravel and stone (Table 6) and were observed throughout field study period of September 1990. The fertilized egg measured 1.18 to 1.87mm. Snow trout up to 200mm length group produced milt and ripe eggs. Snow trout eggs and larvae were collected by fine-meshed nets together with planktons. However, nursery extended from 0.5 to 1.8m depth in certain places. Eggs were collected attached to algae but most of snow-trout larvae were collected free along the sand and gravel bottom nursery. During September many nursery beds had bankful water and supported plenty microscopic plants and animals. These nursery beds support very little fish food after dry spell of summer.

Spawning ground of Point-Nosed Snow Trout or *Chuche Asla*

Spawning ground of the point-nosed snow trout were observed in Barun, Leksuwa and Wang Khola. Good gravel spawning ground of *chuche* were observed in Brun river during September. Fertilized egg of *chuche Asla* were collected from the gravel-bed spawning ground and their larvae and fry were also collected and identified. Normally the *Chuche Asla* breed once in a year; the breeding peak is August-September. During dry season sampling many stranded fingerlings of *Asla* were observed. Therefore, it is necessary to maintain considerable flow in the dewatered section of the downstream section.

Table 5 : Status, Abundance, Migratory Trend and Economic Importance of Important Fish Species in Arun River

Name of the species	*Locality*	*Status*	*Composition schools Size*	*Economic value*
1. *Schizothorax progastus*	I-VI	Local migrant	55-857	Excellent food fish
2. *Schizothorax richardsonii*	I-I	Local migrant	55-187	Delicious food fish
3. *Schizothorax* plagio*stomus*	I-V	Local migrant	80-1000	Excellent food fish
4. *Garra gotyla*	I-VI	Resident	40-50	Largely eaten
5. *Noemacheilus beavani*	I-VI-III	Resident	30-200	Eaten by low castes
6. *Acrossocheilus hexagonolepis*	II-VI	Local migrant	50-90	Highly prized game fish
7. *Barbus chilinoides*	III-IV	Migrant		Good for eating
8. *Puntius conchonius*	III-VI	Resident	20-30	Sun dried for making condiment
9. *Barilius bendelisis*	III-VI	Resident	20-200	"
10. *Barilius barna*	I-III	Resident	5-8	"
11. *Glyptothorax cavia*	III	Migratory	3-5 8-10	Very oily, used for making fish oil
12. *Glyptothorax trilineatus*	I-VI	Local migrant		"
13. *G. t elchitta*		Local migrant		"
14. *G. pectinopterus*	I-VI	Migrant	3-5	"
15. *Pseudecheneis sulcatus*	I-VI	Migratory	3-5	"
16. *Euchiloglanis hodgarti*		Resident	3-100	Used in gastric aliment and hyperacidity
17. *Psilorhynchus pseudecheinus*	I-VI I-VI	Resident		Used in gastric aliment and hyperacidity
18. *Heteropneustes fossils*	III-VII	Resident migratory over land locally	10-20	Poorly eaten, used for convalescent patient

(contd.)

Name of the species	Locality	Status	Composition schools Size	Economic value
19. *Channa punctatus*	III-VII	Resident	20-200	Poorly eatem, used for convalescent patient.
20. *Channa gachua*	III-VII	Resident migrate over land locally	15-200	"
21. *Amphipnous cuchia*	III-VII		8-10	"
22. *Anguilla bengalensis*	IV-VI	Large scale migrant	8-50	Highly prized game fish
23. *Macrognathus aculeatus*	IV-VI	Resident	3-15	Excellent food fish
24. *Mastacembelus armatus*	IV-VI	Resident	3-8	"
25. *Xenentodon cancila*	IV-VI	Local migrant	8-15	"
26. *Amblyceps mangois*	II-VI	Resident	3-5	Used in folk medicine

Table 6: Spawning Habitat by Locality

Locality	*Spawning habitat*	*Status, abundance and migratory behaviour of fishes*
Locality I : The confluence of Arun with Thudam Khola	Thudam Khola upstream of the dam site is a small torrential stream broken by pools and rapids. The stream is composed of boulder, rubble, cobble and gravel. The pH range 6-7 and current speed ranges 12 to 18°C, oxygen level was 10-15 ppm.	Only local migrants such as snow trout, stone-cats- stone loaches, stone carps were found. They congregate in thousands in clear waters of streams and springs adjoining the sections.
Locality II : Confluence of Arun with Honkong Khola	River is encased and sandwiched into narrow area with good hydrobiological condition.The river bed is composed of fine sand. Temperature of water ranges from 15° to 18°C. Here current speed is moderate, usually 3.0 to 4m/Sec.	In this section migratory spawn of snow trout observed at morning and early dawn.Spectacular migratory runs of snow trout and sucker head *Garra gotyla* were observed.
Locality III : Confluence of Arun with Teji Khola	River narrows here. Hydrobiological condition of river good. Arun river at Teji Khola may be regarded as very ideal for spawning	In this section, migratory runs of pointed snouted snow trout were observed in the evening
Locality IV : Confluence of Arun with Sirsung Khola	Hydrobiological condition is good as in Section III. Good in fish sheltering rock stones, and crevices are abundant. There is very good spawning ground for pointed snouted *Asla*.	Snow trout and rock carp were seen utilising gravel beds of Sirsung Khola
Locality V : Confluence of Arun with Barun Khola	River widens. The temperature ranges form 15 to 22°C and current speed 0.3 to 2.9m/Sec. River is boulder strewn. Rich habitat for *Katla*.	A few*Katle* were seen migrating upstream. However, snow trouts utilized all spawning ground of the area. A few larvae of snow trout, *Katle* were sampled in this section.
Locality VI : Confluence of Arun and Leksuwa	Arun river widens, create good spawning habitat for majority of migrants. Spawning assembly were on the banks of pools and rapids. They were very shallow with natural depression, easy for deposition of eggs. Spawning ground was dominated with rocks, pebble, fine sand. The current velocity of the spawning ground was 1-3m/Sec, depth 1 to 3m and pH 6-7 and gravel size 100 to 200mm.Inter-gravel waterflow conditions were ideal for gravel spawners suc as snow trout and *Katla*	Spectacular migration of pointed snouted snow trout.
Locality VII: Confluence with Iksuwa Khola Arun III site	River widens and joins many feeder streams. Spawning beds at Iksuwa, Basuwa are ideal for *Katla* and snow trout. Spectacular upstream migration was observed. Spawning assembly composed of rock, boulder gravel bed with fine sand. The current velocity was 1.5 to 3.0m/Sec., depth 2-5 metre, gravel size 20 to 150 mm. Spawns of *Katla* were detected here.	Migratory *Katla* and *Thed* were observed in the spawning beds of Iksuwa Khola. Swarms of these fast moving upstream.

Spawning Ground of *Katlae*

Spawning ground of *katlae* or copper mahseer were also noticed in downstream section Iksuwa, Thado Khola and Wang Khola. Spawning ground of *katlae* fish is best represented below Uwa Gaun. These sections of spawning ground, the structure and composition of spawning substrate noted to be the same as described in snow trout because these two species share same habitat. The fertilized egg of copper mahseer measured 2.25 mm. Copper mahseer up to 150 mm produced milt and egg for breeding purpose.

Fig. 5 : Snow trout *Schizothorax plagiostomus*

Fish Exploitation

The people of the Upper Arun do not regularly eat fish. Fishing is not very productive due to the high gradient and low temperature. Due to boulders, there is little opportunity for net casting and wading in the stream sections. Occasionally people of Chepuwa divert the Chepuwa stream where it joins the Arun river and set gradient fish traps. Spindle fish traps are set at the end of gradient trap in the evening and fish are collected in the morning. Only 2-3 kg. of fish were trapped in 12 hours. Most of the people living in Chepuwa are Buddhist and do not fish. In some tea-shops near Hatiya, Shyaksila and Gola, smoked or dried fish are sold to travellers.

Aquatic Invertebrates

Micro-invertebrates (Zooplankton)

There are few species of Zooplankton (5-8) upstream of the proposed dam site. The zooplankton in the Chepuwa Khola consisted of Protozoans, Rotifers, Copepods and Ostracods.

i. *Protozoa* : The population of Protozoa is sparsely distributed in the Upper Arun. The various genera of protozoans recorded in order of abundance were *Vorticella, Actinobolus, Nebalia, Actinophyrs, Stylonychia* and *Diffugia*.

ii. *Rotifers* : The most common Rotifer genera included *Brachionus, Keratella, Filinia, Polyarthra* and *Pedalion. Brachionus* and *Keratella* were dominant in all sections of the river. Other rotifers such as *Filinia, Polyarthra* and *Pedalion* were sparse. A few *Ploima, Monostyla* and *Asplanchna* were also recorded.

iii. *Cladocera* : A relatively low abundance of Cladocerans was recorded in the Arun River. Both upstream and downstream of the dam, numerous genera of *Cladocera* were found, e.g. *Diaphanosoma, Simocephalus, Daphnia, Ceriodaphania, Moina* and *Alona.*

iv. *Ostracoda* : Ostracods are sparsely distributed in the project area and appear to be only represented by the genus *Cypris.*

v. *Amphipoda* : Amphipods (*Gammarus pulex*) are present in the Gola section.

Micro-Invertebrates (Molluscs, Insects and Crustaceans)

The river section above the Barun confluence has a high current velocity and submerged vegetation is sparse. Sampling revealed that of the total micro-invertebrate fauna, molluscs and insects were 46.5 and 33.8 per cent respectively.

At the Arun-Leksuwa Khola confluence, *Corixa* is the dominant genus with 21 per square meter. Next is *Hypoporus* with 11 per square meter. Besides these, *Ephemera* (nymph), *Hexagenia* and *Anax* were also abundant. The molluscan population was represented by *Lymnea, Gyrenxus* sp., *Corbicula* sp. *Penaeus indicus* is the most common decapod with a total of 13.8 of macro-invertebrate population. The list of the aquatic insects is given in Table 7.

The average standing crop of macro-invertebrate biota at the Arun-Leksuwa Khola confluence during December is estimated to be 200 units/m^2 by number and 0.5 kg/m^2 by weight. Macrofauna at the bottom of pools consisted of oligochaeta (*Tubifex* sp. *Eropdella*), Crustacea (*Leander* sp.) Insecta : *Ephemeroptera* (*Pentagonia* sp.) Odonata (*Aphylia* and *Perithemus* sp. Coleopteran larvae, Diptera-*Chaoborus* sp. and *Tendepes* sp., *Heterodonta, Corbicula* etc. Fauna at the bottom of the reservoir section was dominated by insects including chironomids (52%) followed by molluscs (31.4%), oligochaets (15.8%).

Sampling suggested that insects preferred rapids zone at a depth of 1 to 3m. while molluscs preferred a depth of 4 to 8 m. The standing crop of macro-invertebrate biota in the downstream section is not much different from that of the proposed reservoir area. Insect production at the dam site is about 5 gm/m2 which is optimum for sustaining food chain for coldwater fish in the area.

The present productivity of fish has been estimated on the basis of interviews with the local fishermen and fishing contractors. There are around 250 professional fishermen in the Upper Arun near Sebrung, Hatiya and Gola villages. Based on interviews with local fishermen, the annual catch is estimated to be approximately 17 kg/ha. of river surface. The total annual fish catch from the whole upper Arun region is estimated as about 8-10 metric tons.

The fishing season in the Arun river starts in September - October and continues until the end of May. Fishing is most extensive after January-February. During winter and summer, the gear most commonly used are cast nets, dip nets and hook lines. The mode of fishing in autumn involves the setting of gradient traps.

Local cast net with mesh size 3-8 mm, is used to collect wide variety of fish. Locally made hooks and snare loops are used to catch large sized fish. Fish are caught by diverting stream channels. Poisons such as lime powder of cactus *Euphorbia royleana*, seeds of Timur (*Xanthoxylum alatum*) and Chila (*Cascaria tormentosea*), etc. are also used for fishing.

People make cast nets and gill nets from a local durable fibre plant known as Allo (*Giardinera heterophylla*). The cast nets are fine meshed and last for a year. Various kinds of fish traps are used for fishing in shallow brooks, creeks and diverted channels of the river. The most effective gears are lift nets and dip nets. Besides these, fish traps are operated in rice fields, irrigation channels and shallow rapids of stream. This equipment is used by women and children for catching minor carps, tadpoles, tree frogs, larvae and crabs. Fishermen also use rip hook for catching sport fish. Snowtrout is first jigged and wound by a hook and then caught. Occasionally, fishermen use fish snares or loop (Manim) for capturing live snow trout. Snow trout enter the hollow of self-collapsing nylon loop and get trapped.

Diverting the stream and using poisons is a common fishing practice. Fish are diverted by the force of current and compelled to enter the spindle-shaped bamboo fish traps *(Khungo)* set at the lower end of the stream channel. Fish shelters are also created in the river bed during the late autumn (October - November) to catch fish after their spawning run. The brush barrier set across the river stop fish from migrating and deflect them toward the fishing channel where they are trapped and collected.

Fishing cycles in the upper Arun are closely linked to hydrological condition of the river. In autumn (September to November) fishermen divert forking branches of the river and create fishing channel called *duwalo*. The labour cost for creating a *duwalo* ranges from Rs. 500 to 1000 and takes about one month to prepare.

Fishing usually starts in the morning at about 8 to 10 AM. During the day time, fishermen are engaged in farming. However, some fish in the evening and until mid-night during the full-moon periods. Bamboo gradient *(Tip)* traps are usually set in the evening during February and in October. During high water phase of May and June, fish are caught and sold at Rs. 40-75 per kg. Small fish are smoked and packed in split bamboo basket. On an average, a basket contains 300-500 grams of dry fish and is sold for Rs. 50.

Explosives are the most commonly used illegal method of fishing. Harmful insecticides are also used in creeks and rivulets near the proposed dam site. Fish poison extracted from *Ichthyotoxic* plants are mixed with mud and sand and is released in fish shelters in crevices and pools of diverted stream channels. This is a harmful practice which not only kills adult fish but also the developing eggs and larvae.

Table 7: List of Aquatic Insects in the Arun River

Order	*Family*	*Sub-Family*	*Gents*
Coleoptera	Dytiscidae	-	*Cybister*
	Hydrophillidae	-	*Hydrophilus*
	Halpidae	-	*Halipus*
	Gyrinidae	-	*Gyretes*
	Elmidae	-	*Stenelmis*
	Dytiscidae	-	*Laccophilus*
	Psephenidae	-	*Psephenus*
Plecoptera	Nemmouridae	Taenopterygidae	*Brachyptra*
	Peltoperlidae	-	*Peltoperla*
	Perlidae	Acrinouridae	*Paragentia*
	Perlinae	-	*Nemoura*
Hemiptera	Corxidae	-	*Corixa*
	Corixidae	-	*Micronecta*
	Hydrometridae	Gerridnae	*Hydrometra*
	-	-	*Gerris*
	-	Nepidae	*Limnometna*
			Nepa
			Ranatra
	-	Notonectidae	*Anisops*
	-	Pleidae	*Pelopea*
Trichoptera	Leptoceridae	-	*Leptocerus*
	Psychomyiidae	-	*Psychomyia*
		-	*Parasetodus*
Lepidoptera	Pyralididae	-	*Nymphula*
Diptera	Nematocera	Simulidae	*Simulium*
	-	-	*Tendipes*
	-	Psychodidae	*Pericoma*
		Tipulidae	*Tipula*
Odonata	Aeschnidae	-	*Anax*
	Gomphidae	-	*Octogomphus*
Zygoptera	Lestidae	-	*Lestes*
Zygoptera	Agricnidae	-	*Hyponeura*
Collembola	Hypogastruridae	-	*Hypogastrura*
Ephemeroptera		Baetidae	*Baetis*
			Himalayana;
			Rhithrogena
		Caenidae	*Caenis*
		Ephemerellidae	*Ephemerella*
		Heptagenidae	*Epeorus, Iron*

CONCLUSIONS AND RECOMMENDATIONS

The population of small scale migratory fish such as snow- trout will be depleted owing to the barrier effects of the dam. Provision of fish ladder is not advisable for it is not economically feasible for the free migration of the fish. Therefore, population in the dam site population is to be maintained by constructing fish hatchery, or introducing snow trout fries from downstream areas.

It is essential to have coldwater fish hatchery which will be responsible for the better management of the proposed cold water fishery regulating the catch, conducting induced breeding for coldwater fish as well as supervising coldwater recreational fisheries. This station would have to monitor water quality and conduct regular studies on the limnology of the river so that necessary steps can be taken to maximize fisheries production through proper management of fisheries resources. The area needed for establishing such a station will be 10-25 hectares which may be available near Arun-Barun confluence. About 20 per cent of the area may be used for maintenance of brood stock of snow trout and 75 per cent of the land may be utilized for the production of fry and fingerlings. To supplement fish protein in the diet of the local people culture activity should be intensified in the area where considerable water is available for fish.

The longer-term effects of the project on downstream fish fauna is not possible to predict at this time; therefore, they are to be monitored time and again.

The life histories of important fish-species e.g. copper mahseer and catfishes, to be studied in detail so as to formulate conservation policies.

Introduction of natural fish stocks occurring in lower section of Arun, e.g. copper mahseer *Acrossocheilus hexagonolepis*, should be attempted to develop recreational fishery in the project area.

Fish population existing in the dam site will be depleted due to the barrier effects of dam and dewatering effects of water release in dry period is essential. If the construction of the fish hatchery is considered to be uneconomical, alternatively snow trout population existing in some other river near Arun has to be introduced. For this purpose the river-caught snow trout should be oxygen packed and introduced safely into the reservoir. For long-term conservation of snow trout species, it is highly essential to establish a coldwater fish hatchery near the project area.

Since proposed upper Arun power dam section does not harbor any endangered fish species, the creation of the dam is highly desirable for the generation of power.

It is concluded that the spawning migration of snow trouts, copper mahseer,cat fishes and loaches is timed by season and external factors of environment such as temperature, flow, oxygen, pH, dissolved solids as well as internal environment of the fish. The spawning urge seems to be controlled by the internal physiological rhythm that means site specific visual factors, smell. Two to three groups of eggs were seen in mature female fish, which gives inference that the fishes of Arun are the fractional spawners and spawning takes place after each batch of eggs matured. The spawning season detected to be long, spawning over three months. False spawning was observed first before real spawning took place.

Arun river harbour 26 species of fish which are of great zoo- geographical interest. Both fishes having Pan-oriental and Indo-Chinese affinites were detected. Presence of new species novel to science is possible but needs long term study to detect their presence. During winter and summer sampling period no rare species endemic to Arun were detected.

Arun river does not have any endangered species. No long range migrants were detected during the sampling period. Snow trout and copper mahseer are short-distance migrants; therefore a fish ladder is not justifiable for them. Therefore, impact of project to migratory fish population will be minimal because most of the fish migrate during high water phase of the river and spawning and nursery area will be available for them after establishment of dam. However, attempt should be made to launch a long term habitat enhancement programme at downstream area of Chepuwa dam site to compensate for habitat loss due to dam.

REFERENCES

Bhatta, D. D. and Shrestha, T. K. (1977): *Environment of Suklaphanta*. Curriculum Development Centre. Tribhuvan University, Kathmandu, Nepal.

Eschmeyer, R.W. and Manges, D.E., 1945: Fish migrations into the Norrism Dam in 1943. *Journal of the Tennesse Academy of Science*, Vol 29, No. 1, pp. 92-97.

Frasher, J. C. 1972: Regulated stream discharge for fish and other aquatic resources. *FAO Tech. Paper* No. 112, Rome.

Fox, J. L. 1974: *An ecological survey of proposed Langtang National Park*, HMG, Kathmandu, Nepal.

Hammad, H. Y. 1972: River bed degradation after closure of dams. *J. Hydraulic Div. A. Soc. Civ. Eng.* April 1972. 591-605.

Jayaram, K. C. 1981: *The fresh water fishes of India, Pakistan, Bangladesh, Burma, Sri Lanka: A handbook.* Zoological survey of India.

Lagler, K. F. (ed.) 1969 : *Man made lake planning and development*, FAO, Rome.

Lagler, K. E. 1976: *Fresh water fishery biology.* WMC Brown Company, IOWA.

Leitritz, E. and Lewis, R. C. 1976 : Trout and salmon culture. *Fish Bulletin* 164 State of California, Department of Fish and Game. Sacramento USA.

Low-McConnel, R. H. (ed.) 1966: *Man made lake.* Academic Press, London.

Milhouse, R.T., Wegner, D.L. and Waddle, T. 1984: Users guide to the physical habitat simulation system (PHASIM) *Information Paper* No. 11.FWS/OBS -81/43. U.S. Department of Interior, Fish and Wildlife Service, Fort Collins, Colo.

Mills, D. (ed.), 1990: Strategies for the rehabilitation of Salmon rivers. *Proceeding of a joint conference held at Linnean Society* in 1990. The Chameleon Press Ltd., London.

Nestler, J.M., Walburg, C.H., Novotny, J.F., Jawbs, K.C. and Swink, W.D.1986: *Hand book on reservoir releases for fisheries and environmental quality.* U.S. Army Engineer Waterways Experiment Station Environmental Laboratory, Vieksburg, Mississippi.

NORAD, 1988 : *Environmental Impact Assessment* (EIA) *of development aid projects.* Norwegian Ministry of Development Cooperation. NORAD OSLO.

Ruggles, C.P.1989: *A critical review fish exclusion and diversion from hydroelectric turbines, with special reference to Hadley Falls Station on Connecticut River.* Mononco Maritimes Ltd. Halifax, N. S. Canada.

Shrestha, T.K. 1978: A new report on the induced breeding of Nepalese hill stream trout *Schizothorax plagiostomus*, *J. Ins. Sc.* 1; 171-176.

Shrestha, T.K. 1979: Life history of the hill stream trout *Schizothorax plagiostomus* (Heckel). *J. Ins. of Sc.* Vol. 2, No. 1; 209-222.

Shrestha, T.K. 1979: Studies on resource, biology and ecology of fresh waters of Kathmandu valley with particular reference to fish production, management, marketing and conservation. Research project sponsored by Research Division of Tribhuvan University, Kathmandu, Nepal.

Shrestha, T.K. 1980: The biological indicator of pollution in the river Bagmati. *Tribhuvan University Journal*, XI. No. 1. 117-135.

Shrestha, T.K. 1986: Development of inland fisheries in Nepal. *Nature Conservation Society Newsletter* No. 29.

Shrestha, T.K. 1990: *Resource Ecology of the Himalayan Waters.* Curriculum Development Centre. Tribhuvan University, Kirtipur Campus, Kathmandu, Nepal.

Shrestha, T.K. 1990 : Rare Fishes of Himalayan Waters. *Journal of Fish Biology.* 37 Suppliment. Academic Press, London.

Terashima, 1984 : Three species of cyprinid genus *Schizothorax* from Lake Rara North Western Nepal. *Jap. J. Ichthyology*, 31(2) : 122-135.

Rec. Adv. Fish Ecol. Limn. Eco-conserv.
III :27-31

2

Physical Ecology of Four Lakes of Kumaun : Nainital, Naukuchiatal, Bhimtal and Sattal

S.M. DAS

INTRODUCTION

Many large and small lakes in the Kumaun region have long attracted tourists as the 'Lake District' of Kumaun Himalaya, comparable with the 'Lake District' of England and the group of Kashmir lakes. The Kumaun lakes comprise the only group of mountain lakes in the outer Himalayan region.

The four Kumaun lakes Nainital, Bhimtal, Naukuchiatal and Sattal only had never been worked out for their physical, chemical and biological ecology except for some data for Nainital in DST Project. Their physical ecology is reported in the present contribution. The other lakes viz. Punatal, Malvatal Sariatal and Khurpatal still remain to be fully investigated. All these lakes lie in a belt 24 kms. long and 5.2 kms. broad, which is a mature dissected peneplain lying at an altitude ranging from 1220 m to 1938 m.

The origin of these lakes has been the subject of much speculation and controversy. The most accepted view today is that of Thomas (1952) and Kharakwal (1971), which hypothesises :

(1) an intermontane peneplain was uplifted, eroded by torrential rainfall and streams and then blocked by giant landslips ; and

(2) a belt of incompetent slates had slumped irregularly during the later tensional phase of the uplift of the sub-Himalayan zone. All recent authors agree that there is no evidence of glacial origin of these lakes as suggested by Blanford (1877), the lake rocks' markings hardly dating back to the glacial period (Gansser 1964).

The overall evidence appears to show that the Nainital group of lakes (Sariatal, Khurpatal and Nainital) originated separately from the Bhimtal lake group (Bhimtal, Naukuchiatal, Sattal with Garurtal, Harishtal, Mahratal and Punatal).

RESULTS AND DISCUSSION

An overview of Tables I, II, III and IV will show that the four lakes differ radically in their physical ecology (altitudinal distribution, catchment area, lake area, lake depth, slope zone areas, annual rainfall, Secchi transparency, thermocline, lake colour and lake type).

The largest in area is Bhimtal (63.25 ha.) while the smallest is Sattal I (Garurtal, 17.0 ha.). The catchment area of Naukuchiatal is, however, the largest (18.4 sq. km.), while that of Nainital is much smaller (11.8 sq. km.), followed by Bhimtal (10.8 sq. km.) and Sattal (14.5

sq. km.). Although in surface area Bhimtal is the largest (63.25 ha.), the deepest is Naukuchiatal (40 m at centre), followed by Sattal (30 m) and Nainital (24.2 m), while the shallowest is the large Bhimtal lake (18.0 m).

Both Nainital and Bhimtal have been subjected to heavy silting. Nainital has become shallower by 4 metres in the last 60 years (Neville 1922 recorded its depth as 28 m), while Bhimtal has been converted into a sandy marsh at its northern Mallital (upper) end, with a depth of 50 cms. to 2 m only, affording heavy growth of sedges *(Typha, Phragmites)*.

The slope Zones of the lake basins show a wide variation in low slope and high slope areas. At 0° - 14° slope, the largest area is of Bhimtal (630.0 ha. due to gently sloping hills), while the smallest area lies in Sattal environs (17.0 ha.). Similarly in the 15°-35° category the area is highest in Bhimtal (325.0 ha.), followed by Nainital (177.5 ha.); but even Sattal has 148.0 ha. as 15° - 35° slope. The high slope areas (36°-70°) are largest in Nainital (249.8 ha.) and smallest in Naukuchiatal (11.0 ha.), the high slope areas of Nainital being due to the surrounding high hills and peaks of China, Deopata, Alma, Sherkadanda, Valehead, Handibandi, Kaliakhan and Tiffintop, where barren stony blanks account for as much as 14.8 ha. in the high slopes.

Table -I : *The slope zones of Nainital Basin* (*Catchment area*)

Slopes in degrees	*Area in 00 sq.mtrs.*	*Plant area in 000 sq. mtrs*			
		grass	*scrub*	*conifers*	*oaks*
(a) 0° - 14°	572.33	117.9	78.41	14.88	49.79
(b) 15° - 35°	1775.13	852.0	72.78	63.91	94.68
(c) 36° - 70°	2498.74	299.85	362.32	454.77	984.51
(d) 71° - 90°	143.80	(Stony blanks - 143.80)			

Table - II : Landuse and Vegetation Areas (Hectares)

Data	*Nainital*	*Bhimtal*	*Naukuchia*	*Sattal*
Cultivation	145.1	842.0	40.0	69.0
Grassy Banks	23.0	61.0	13.0	4.0
Scrubland	53.4	70.0	12.0	9.0
Oaks	112.8	23.0	14.8	52.0
Conifers	53.3	52.0	5.2	156.0
Deciduous	12.1	39.0	41.0	100.0
Stony blanks	14.8	-	-	-

The annual rainfall, although heavy in all four lakes, shows a steady decline from high altitude Nainital (2480 mm) through Naukuchiatal (2424 mm) to Bhimtal (2143 mm) and Sattal (2034 mm). The air temperature is lowest in Nainital in winter (average 3°C), but highest in summer (34.5°C). The other 3 lake basins have an average winter temperature of 10.5°C and 32.5°C in summer. The higher summer temperature of Nainital is due not only to buildings congestion and population, but also to the region being a pocket between high hills, aiding retention of solar heat.

The water temperature however shows the altitudinal gradation expected from the 1938 masl Nainital (23.5°C summer, 8.0°C winter) to the lowest (1370 masl) Bhimtal (26.5°C summer; 10.5°C winter), the highest water temperature recorded being in the lowest lake Naukuchia Tal (28.0°C summer; 10.5°C winter) with an altitude of 1230 masl. The minimum

Table III : Physical and Chemcial Parameters of Polluted Waters of Eutrophic Lake Nainital (Monthwise Average Value)

Month	Temperature (°C)	Colour	Odour	Turbidity (cms)	PH	Dissolved 02(ppm)	Free Co_2 (ppm)	NH_4 mg/L	Alkalinity
January	9.5	Brownish green	Fishy	170	6.8	1.5(0.8)	28.0	0.28	180
February	9.5	Brownish green	Fishy + Methane	160	6.8	1.54	20.0	0.28	180
March	15.0	Yellowish green	Methane	160	7.5	4.8	22.0	0.25	184
April	18.5	Pale green	Fishy + Methane	135	7.8	5.10	26.0	0.25	186
May	22.0	Dark green	Strong Methane	130	8.0	5.12	22.0	-	190
June	20.0	Dark green	Methane	130	8.0	5.12	22.0	-	192
July	23.5	Yellowish green	Weak Methane	108	8.5	6.8	14.0	-	192
August	23.2	Yellowish green	Weak Methane	100	8.8	8.2	4.0	-	240
September	23.5	Yellowish green	Fishy + Methane	122	9.2	9.4	10.5	-	258
October	23.0	Dark green	Fishy +Methane	140	9.0	10.2	15.2	-	312
November	16.5	Bluish green	Fishy	160	8.8	8.4	20.0	-	294
December	7.0	Brownish green	Fishy	178	8.6	4(1.9)	30.0	-	350

Table -IV : Hydrological Data of Oligotrophic Bhimtal Lake (Monthwise and Average Value) (1978-79)

Month	Temperature °C	Turbidity (cms)	Water colour	pH	Dossolved O_2 (ppm)	CO_2 (ppm)	Total alkalinity ppm	Total Hardness ppm
1978								
April	23.5	430	Bluish green	7.5	10.0	-	60.0	45.0
May	24.0	410	Green	7.5	6.8	-	56.7	48.0
June	26.0	420	Bluish green	8.2	5.7	-	40.2	48.0
July	23.0	385	Bluish green	8.5	8.8	4.0	35.5	42.5
August	23.5	400	Yellowish green	8 5	9.2	4.0	34.0	40.5
September	22.0	435	Bluish green	8.5	9.2	4.0	31.5	38.0
October	20.0	465	Light green	8.5	10.8	-	32.0	39.0
November	16.0	460	Dark green	8.0	13.6	-	29.0	40.0
December	10.0	500	Green	8.2	13.3	3.0	26.6	42.0
1979								
January	12.2	490	Light green	8.1	13.0	3.0	31.5	43.5
February	16.0	460	Dark green	7.8	13.1	2.0	37.5	43.5
March	17.7	445	Bluish green	7.5	16.8	-	43.7	44.0

winter temperature thus ranges between 8.0°C (Nainital) to 10.5°C (other lakes). This belies the common view that mass mortality of fish in lake Nainital occurs every winter due to cold. A difference of 2°C can not cause mortality, since there is no fish mortality ever in Bhimtal, Naukuchiatal and Sattal.

Secchi disc transparency readings clearly delineate the 4 lakes in to clear water (oligotrophic) and turbid water (eutrophic) types. Lake Nainital has only 1.5 m transparency, while the other 3 have values of 5 m or above. This clearly shows Nainital to be eutrophic (established by chemical and biological parameters also (Das, 1978, 1979, 1980).

A thermocline has been found in summer in all the four lakes. Nainital and Naukuchiatal have an epilimnion of 6 m and 8 m respectively; while Bhimtal and Sattal have thermocline at 3.5 m and 4.2 m respectively. Temperature differences are high between epilimnion and hypolimnion, being as much as 14°C in Nainital and Naukuchiatal lakes.

Ecological data on vegetation and landuse also exhibit wide differences in the catchment areas of the 4 lakes. Expressed in hectares, the cultivated area is highest in Bhimtal basin (842.0 ha.), and lowest in Naukuchiatal (40.0 ha.) . Grassy banks are largest in Bhimtal (61.0 ha.) and smallest (4.0 ha.) in Sattal. Scrub and shrub land again is the highest in Bhimtal (70.0 ha). and lowest in Sattal (9.0 ha.).

As regards forest stands, oaks occupy the largest area in Nainital (112.8 ha.), while the smallest area is for Naukuchia tal (14.8 ha.), which is accounted for by the high altitude of Nainital (Oak treeline) and the high rainfall. Conifer area is highest in Sattal (156.0 ha.) and lowest in Naukuchiatal (5.2 ha.) ; while deciduous tree areas are also highest in Sattal (100 ha.). These large stands of conifers and deciduous trees in Sattal are presumably due to its comparative inaccessibility to man and less denudation by tree felling. Oaks bind the soil through spreading roots, while conifers do not; which accounts for the high landslides and hillslides in regions lacking oaks (such as the major 1981 hillslide near Waldiakhan on Nainital motor route).

CONCLUSION

The present ecological study throws new light on the nature and origin of Kumaun lakes and the difference in their physico-ecological parameters. As the eutrophic lake Nainital and its environs are now polluted and congested due to population explosion and tourism, attempts should be made to shift the population expansion to the oligotrophic lake environs of Bhimtal, Naukuchiatal and Sattal lake basins, which are still in their pristine natural condition, and thus create a greater Nainital of the future.

REFERENCES

Blanford, H.F. (1877) : On the glacial origin of Kumaun Lakes. *Proc. Asiatic Soc. Bengal 3* (1) : 63 -72.

Das, S.M. (1978) : Ecology and Management of Freshwaters in highlands of India. *Proc.DST Seminar on Development in Himalayan region. Govt of India pubs : 256 - 271.*

Das, S.M. & L.S. Khanka (1979) : Physiography and Hydrography of lake Nainital. *Geogr. Review, India, 41* (2) : 139-143.

Das, S.M. & J, Pande (1978) : Physico-Chemical and Biological Pollution in lake Nainital. *Indian J. Ecology, 5* (1) : 7 - 16.

Das, S.M. & J. Pande (1979) : Effect of increased landuse upon freshwaters in Central Himalayas. *Progr. Wat. Tech. London,* 1979; *11* (6) : 195 - 208.

Gansser, A (1964) : *The Kumaun Himalayas.* Wiley Interscience, London.

Kharakwal S.C. (1971) : Origin of Kumaun Lakes. *J. Wadia Instt. Him. Geology* : 147 - 152.

Thomas, A. N. (1952) : The Nature and Origin of Kumaun Lakes. *Geol. Mag. 89* (6): 386 - 400.

Rec. Adv. Fish Ecol. Limn. Eco-conserv.
III :33-37

3

On The Discovery of an Additional Component in the Weberian Chain of Ossicles in two Air-Breathing Catfishes *Clarias* and *Heteropneustes**

*D. K. MASSEY***

INTRODUCTION

The study of the Weberian apparatus presents an intricate problem due to their small size, internal location, their role in fish systematics and homology. Not only have different researchers given different views on the homology of the Weberian apparatus, but different views have also been expressed as regards their importance in fish systematics.

WEBER (1820) first discovered the existence of movable ossicles between the auditory organs and the gas bladder. This interconnected chain of ossicles was homologized by him with the ear ossicles of mammals and man and he named them malleus, incus and stapes. Later workers named these as Weberian ossicles after *WEBER* and showed that they were in reality four in number and are modifications of the first four vertebrae.

The main workers abroad in this field of morphology have been *WRIGHT* (1884) *SAGEMEHL* (1891), BRIDGE and *HADDON* (1893), *NUSBAUM* (1908), *GOODRICH* (1909, 1920), KINDRED (1919), *MATVIEW* (1929), *NELSON* (1948), and *ALEXANDER* (1962, 1964).

SARABAHI (1932) gave an account of the Weberian ossicels of *Labeo rohita*, *DAS* and *MASSEY* (1965) gave a detailed account of the Weberian apparatus of some Indian Ostariophysi.

MATERIAL AND METHODS

Some fish such as *Clarias batrachus* and *Heteropneustes fossilis* could be had in the living condition from the local market. For dissection of Weberian ossicles 24 hour fixation in formol-alocohol (10% formalin and 90% alcohol 1 : 1) and later preservation in 6% formalin yielded good results. By this method the ligaments become hardened enough to be taken out intact. Bulk staining *before dissection revealed an* additional component of the Weberian chain in *Clarias* and *Heteropneustes.*

---

* Ph. D. Thesis : "Studies on comparative functional anatomy of Weberian Apparatus and Gas Bladder in some Teleost fishes". Kanpur University, 1981.

** Present address of the author : 87 Dahilamau, Pratapgarh (U.P.)-230001, India.

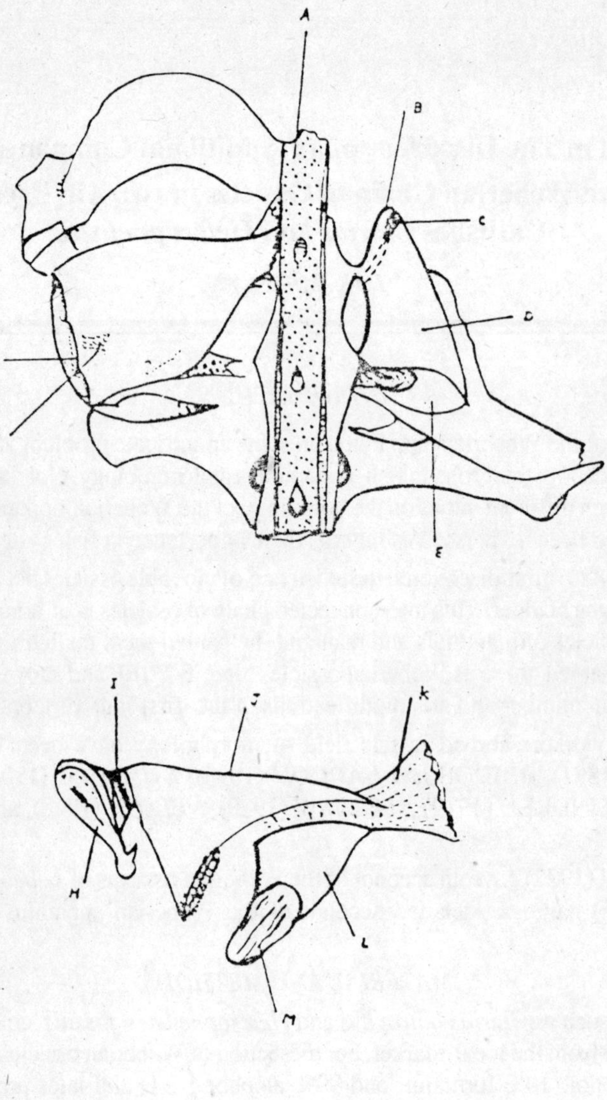

Fig.1 : Weberian Apparatus of *Heteropneustes fossilis*

A - Vertebral Column
B - Scaphium
C - Intercalarium
D - Tripus
E - Accessory Component
F - Lateral Ossification
G - Bladder
H - Scaphium
I - Intercalarium
J - Tripus
K - Bladder
L - Accessory Component
M - Ossa suspensorium

COMPARATIVE MORPHOLOGY

The Weberian apparatus usually consists of the modified first four vertebrae (complex vertebrae) associated with the paired chain of Weberian ossicles and gas bladder (pars auditum) at the front end.

Heteropneustes fossilis and *Clarias batrachus* have an additional movable bony element, the accessory component, attached to the posterior ramus (transformator process) of the tripus, reported here for the first time, and named as the "accessorium" by the present author.

In *H. fossilis* the accessorium is joined suturally to the inner surface of each transformator process. The other end of the Accessorium is then attached to the body of the complex vertebra by tough ligaments *(Fig. 1)*. In *C. batrachus* the accessorium is a large elongated bone nearly equal to the tripus itself, joined to the tip of the tripus by ligaments at the middle. The outer end of accessorium attaches the bladder fibres, while the inner or lower end is attched to the body of the complex vertebra by strong ligaments *(Fig. 2)*.

TABLE

	Name of fish	*Length of fish cm*	*Length of Scaphium mm*	*Length of Tripus mm*	*Length of Accessorium mm*
1.	*Heteropneustes fossilis*	20.0	1.35	4.15	3.25
2.	*Clarias batrachus*	22.0	2.6	5.5	4.00

MECHANISM OF WORKING OF THE WEBERIAN OSSICLES

In the fishes under study there seems to be a relationship between the structure and special features of Weberian apparatus of Ostariophysi (Carps and Catfish) and their ecological niche. For instance the bottom dwellers with sluggish movement or sedentary life and dorso-ventrally flattened (depressed) bodies (e.g.*Bagarius bagarius*, *Clarias batrachus* and *Heteropneustes fossilis*) the gas bladder is enclosed in a bony tube, reduced in size and placed transversely to the body axis. The tripus is situated in a deep groove in the vertebral column and is paper thin and slender. This appears to be an adaptation to high water pressure in these demersal fishes; and both gas bladder and tripus are securely protected.

The present study shows that though the method of transmission of hydrostatic acoustic stimuli by the Weberian ossicles differ in carps and catfishes and even among catfishes themselves, the mechanism of functioning of the apparatus is, by and large, similar. Special modifications in the catfishes with an extra ossicle (*C.batrachus, H.fossilis*) - the accessorium, paper thin tripus, transverse gas bladder and a special deep groove in the vertebral column, are features which have made the whole system extra sensitive and efficient for transmission of vibrations.

In these demersal silurids the reduced transversely-placed gas bladder is enclosed in a bony capsule open on the lateral aspect. The gas bladder is in contact with the skin

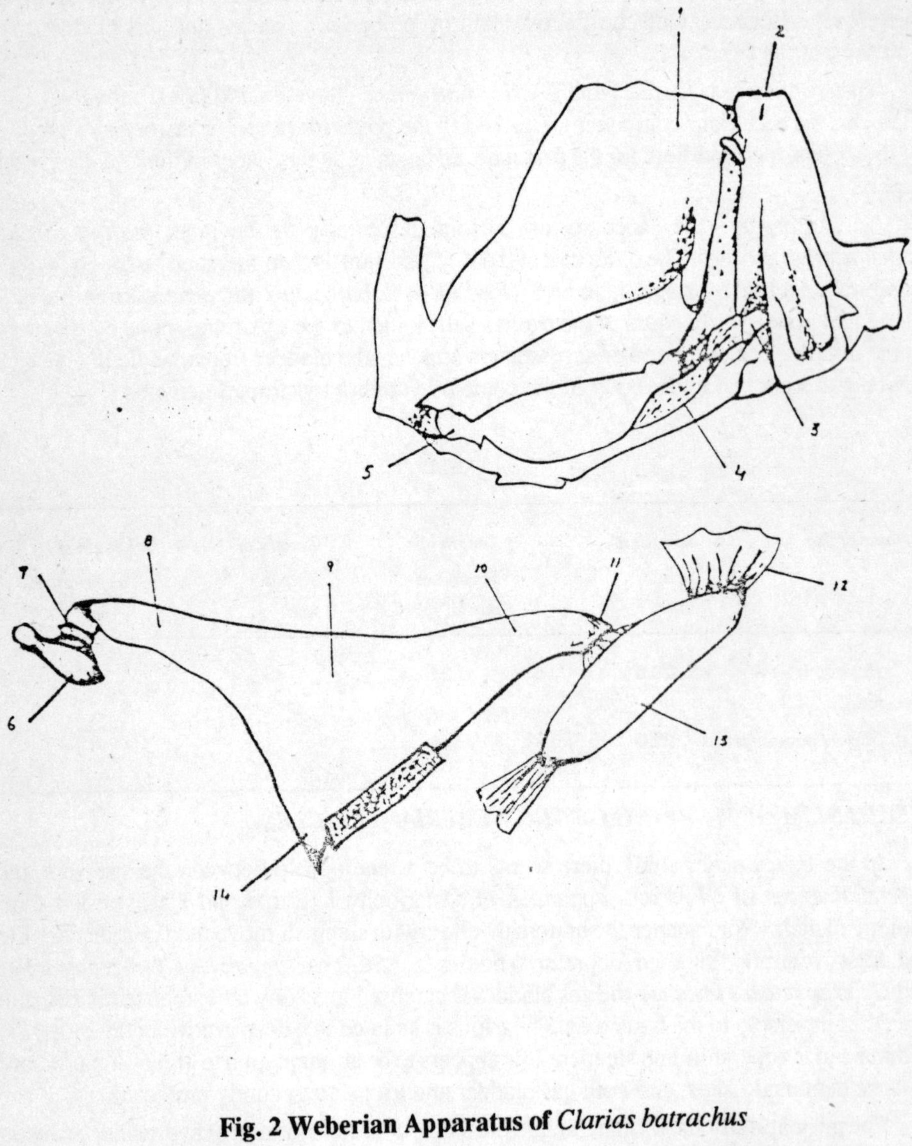

Fig. 2 Weberian Apparatus of *Clarias batrachus*

1. Tripus
2. Scaphium
3. Centrum
4. Accessory Component
5. Bladder
6. Scaphium
7. Intercalarium
8. Anterior ramus
9. Tripus
10. Posterior ramus
11. Transformator process
12. Bladder
13. Accessory Component
14. Articular process

immediately behind (posterior to) the post temporal bone of the skull. Hydrostatic pressure and possibly sound vibrations are felt by this part of the skin pressing the lateral end of the gas bladder. Vibrations are communicated by the latter to the ear. The bony capsule extends laterally as a tube and is in contact with the skin immediately behind the post temporal bone of the head. The small round area of the skin over the openings is seen to be thinner than the surrounding skin, almost like the skin over the tympanum of amphibians and may be considered as a case of convergent evolution. A slight touch or mechanical pressure over this part registers corresponding pressure inside the gas bladder and movement of Weberian ossicles can be seen. Thus it is reasonable to believe that the pressure of the surrounding water can be registered by the small area of the skin just above the gas bladder. This pressure is transferred to the gas bladder, which in turn is transmitted to the Weberian chain through the accessorium and tripus.

The extra component (accessorium) of these two air-breathing catfishes suggests a greater use of the chain, bringing about the ossification of the ligament to form this extra component, giving additional strength to the regular members of the Weberian chain of bones.

These fishes are demersal and bottom feeders but at the same time air-breathing. This means that they have to make frequent trips to the surface for air and also to the bottom for food. The hydrostatic function and also the acoustics have to be of a high order. This probably resulted in the ossification of ossa suspensorium and attaching ligaments to the gas bladder and tripus. The tripus being thin and slender needed a back-up in strength yet maintaining true transmission which has been provided by the accessorium.

Therefore the arrangement is like this :

Cyprinids: Claustrum, scaphium, intercalarium and tripus

Silurids : Scaphium, intercalarium and tripus

Clarias & Heteropneustes : Scaphium, intercalarium, Tripus and accessorium.

REFERENCES

Adams, L.A.(1964) : The structure of Weberian apparatus in Siluri. *Proc. Zool. Soc. London, 143* (3) : 419-440.

Bridge, T.W. & A.C. Haddon (1889) : Contributions to the anatomy of fishes. II the air bladder and Weberian ossicles in the siluridae. *Proc. Roy. Soc. London, 46* : 309-328.

Das, S.M. &D.K.Massey (1965) : On the Weberian appartus of some fresh water Ostariophysi belonging to the families Bagridae, Sisoridae & Cyprinidae. *Ichthyologica,4* (1-2) : 37-43.

1967 : On the cranial osteology and Weberian apparatus of *Clarias batrachus* (Ham.), Siluriformes. *Kashmir Science, 4* (1-2) : 23-37.

Dutta, N.C., A.K.Saha & A. Baidya (1975) : Comparative study of Ostariocranium and Weberian apparatus of *Clarias batrachus* (Linn.) and *Heteropneustes fossilis* (Bloch) (Pisces). *Zool. Anz. 195* (5/6) : 374-386.

Massey, D.K. (1970) : A comparative study of the Weberian apparatus in four Indian teleost fishes (*Clarias batrachus, Heteropneustes fossilis, Eutropiichthyes vacha* and *Ompok pabo*). *Ichthyologica, 10* (1-2) : 16-20.

Mookerjee H.K.,Ganguli, D.N. & Brahma, S.K (1952): On the structures of the Vertebral column and Weberian Ossicles in the Carp minnow, *Esomus danricus. Proc, Zool Soc. .,Bengal*, 5.

Rec. Adv. Fish Ecol Limn. Eco-conserv
III :39-40

4

A Simple Method of Water Purification by Reeds and Rushes

S.M. DAS

It has been estimated that about 80% of the diseases in developing countries (including India) are caused by impure Water. Impure or polluted water in lakes and rivers also causes heavy fish mortality (as in lake Nainital) every winter. This pollution has increased today due to rivers and lakes losing their power of oxygenation and detoxification by reduction or destruction of the ecosystem of reeds and rushes and an increase in pollution by man. The plants die and also cause organic pollution. Their self-cleaning properties are thus getting less and less with every passing year.

The developed countries have built up huge water treatment plants like complex chemical factories, with the latest engineering skills, which are highly expensive. They also sell their plants to developing countries at high cost as a business project.Treatment of polluted water by reeds and rushes is quite inexpensive, hardly 20% to 30% of conventional treatment by chemicals, power plants and large gravel and sand tanks, which often get clogged.

The green plant (reeds and rushes) process is much simpler, cheaper and more lasting. The main plants used in Australia and FRG and available in India are *Typha, Phragmites, Cyperus, Arundo, Scirpus, Jancus, Thysalolema* and *Atriplex*. Other local littoral plants may also be screened. The plants must be cut after the vegetative period and planted on the gravel and sand beds of percolation wells, tanks and water channels, for filtration. The colloids get flocked out by the root ecosystem. This green plant and filter takes only a short time to reduce turbidity, BOD and suspended matter. (In one case turbidity decreased from 0-53 cm, BOD from 1300 to 5.5 and SSM from 8.25 to Zero in less than half an hour.)

Disinfection is also performed by the root ecosystem of these reeds and rushes, which destroy pathogenic bacteria, viruses and eggs of *Ascaris* and other intestinal parasites of Man, so common in Nainital. The biochemical processes of this disinfecting property are still being worked out. *The effluent* from such a green plant-sand gravel filter also gets enriched in oxygen nine to ten times as compared to the *influent*.

Phragmites and *Scirpus* are very efficient in stabilising pH as well (from 4.0 to 12.0, to near 7.0 (neutral)). Another special property of the root system of these reeds and rushes is that the plants absorb and accumulate many heavy metals (as found in Nainital lake water and sediments), such as Cu, Co, Ni, Mn, Cr, Cd and Hg. They store 300 to 400% more heavy metals than in their natural habitat.

Surprisingly this plant root complex grown on gravel sand filters can be adapted to live on 100mg/1 phenol (C_6H_5OH) as also detoxify Xylol, Kresol, Pyridine, etc. (and many other chemicals released by hospitals and laboratories, as in lake Nainital) as reported abroad. Also phosphates, borates and sulphates from washing powders and detergents are reduced by the combined action of green plants and associated micro-organisms in the filters.

Thus the system elaborated above requires : (i) large pits (wells) filled with gravel and sand, into which water is pumped in ; (ii) while the reeds and rushes are planted in the gravel and sand (exposed to sunlight); (iii) the filtered water is carried by under drains into channels with filtration galleries, planted with *Phragmites, Scirpus, Juncus, Iris, Kochia, Atriplex* etc. This biological complex, along with associated micro-organisms, would effect a remarkable detoxification as well as desalination.

The long life of such a system can be envisaged quoting such a treatment site working in Krefeld/FRG over 13 years continually without any problem of maintenance and an output of 400m^3 potable water per hour.

Rec. Adv. Fish. Ecol. Limn. Eco-conserv.
III : 41-45

5

Faunal Survey and Ecology of Fishes of Tehri-Garhwal

S.M.DAS.

Introduction

Physiography of Tehri Garhwal

TOPOGRAPHY : Tehri-Garhwal lies between latitudes 30°3' and 31° 18' N and longitudes 77° 49' and 79° 24' E with an area of 4421 sq. Km. It is bordered in the North by District Uttarkashi, in the East and South by District Pauri-Garhwal and in the West by District Dehra Dun. Tehri lies in the Himalaya and is composed of a tangled series of ridges with innumerable spurs separated by narrow valleys.

The general direction of the main ridges is from N-E to S-W; radiating from a lofty series of mountains, horned peaks, serrated crests of high ridges, cirques and glaciers, snow-clad slopes, hanging valleys, cascades of sparkling water from melting ice and snow, torrential rapids and gigantic escarpments comprise a gorgeous topography. Deep canyons, roaring streamlets and huge boulders present a beautiful scenery and inspiring landscape.

Tehri city is situated on the left bank of river Bhagirathi, between latitudes 30° 22' 54" and longitude 78° 31' 3", at an elevation of 683.4 masl. River Bhilangana joins Bhagirathi a little upstream from the town on the left side.

The rivers Bhagirathi, Bhilangana, Alaknanda, Mandakini, Ganga and Yamuna drain the whole waters of Tehri-Garhwal. This river system provides an excellent example of antecedent drainage.

From the remote corners of this hilly tract, numerous streams and rivulets (locally known as *Gads* and *Gadheras*) drain the area. All the main rivers have a peculiar common tendency to flow for some distance parallel to the mountains, but sooner or later they take an acute bend to flow in deep transverse gorges, hundreds of metres in depth at places.

CLIMATE : Owing to the complicated relief of this mountainous tract, microclimates are of considerable significance. There are not only variations but also complexities of climate and weather, accentuated by the relief of the land. The variations in exposure to sunlight and to rain-bearing winds produce very intricate patterns of local climate. During summer, the valleys have a tropical climate; valley winds in narrow valleys and heavy fog during winter in wide valleys are conspicuous features of the weather in this region. The precipitation at every locality is directly related not only to the altitudinal zone in which it exists, but also to its situation in relation to the ridges or the overlapping spur.

The monsoon starts towards the end of June and ceases by the middle of September. Winter depressions cause snowfall from January to March. April - May experience thunder and occasional hailstorms. In May and the first half of June convectional rain occurs in the afternoon in small amounts (10 to 25 mm.) especially in the valleys practically every third/ fourth day (D.Dun), often even at high altitudes. In general rainfall averages between 37 - 50 cm. from June to September in Outer Himalayas and 20 to 25 cm. in the Inner Himalayas.

The microclimatic conditions differ from place to place and locality to locality according to : (i) the direction of ridges, (ii) degree of slope, (iii) sunny or shady aspect of slope, (iv) intensity of forest cover, and (v) nearness of glacier. The mean annual temperature range is 3°C to 40°C.

The total average rainfall is about 227.9 mm per annum, 85 % of which falls in the months of June, July, August and September. The year may be climatically divided into three parts:

(1) *Dry Summer* : It begins from March to May when the average temperature is 29° - 43.2°C, and extends to mid-June when the maximum temperature sometimes reaches upto 44°.6 C. This period is mainly dry, except for the occasional rains.

(2) *The Monsoon Season* : It commences from the end of June. The SW monsoon winds emerging from the Bay of Bengal result in rains which last for about four months. The temperature ranges between 31.6° C and 44°C , the humidity increases to about 98%. August has the maximum rainfall, sometimes having as much as 806.8 mm. of rain, though the average rainfall for this month is about 722 mm.

(3) *The Winter Season* : It commences from November to February when the average maximum temperature is about 20°C; nearly 0.8 to 45.0 mm of rainfall occurs in the winter also.

The Rivers of Tehri - Garhwal

Yamuna river drains the whole region along the western part of Tehri district and forms a line of demarcation between this and district Uttarkashi. From the east of the district Mandakini river drains the water of that area and forms a line of demarcation between Tehri and Chamoli districts. Alaknanda river separates Tehri from Pauri, from Rudraprayag to Deoprayag (On the southern region).

At Deoprayag the Bhagirathi and Alaknanda merge to form the Ganga. Bhagirathi also forms the line of demarcation between Tehri and Uttarkashi district and from Bhadiyana onwards it flows inside Tehri district.

At Tehri town River Bhilangana meets the Bhagirathi after originating from Khatting glacier and draining the whole Bhilangana valley.

(1) Mandakini : River Mandakini originates at SE base of Kedarnath peak at an altitude of 3583 m., lat. 30° 47' and long. 79°8' E. It flows southwards and is the main river of Kedar valley, with a number of tributaries. Near Tilwara the Lester stream joins the Mandakini on the right bank. It flows southwards and joins River Alaknanda on the right bank at Rudraprayag at an latitude of 30°17' 19" N and longitude 79° I' 32" E. The watershed (parting ridge) between Tehri and Chamoli districts runs from Agastyamuni to Rudraprayag.

(2) Lester stream at Tilwara : It originates from Kinkholakhal at 3400 masl., N. lat. 30° 27' 30" and E. long. 78°55'. It is a large torrential stream surging through rock gorges and narrow valleys. The stream has many lips, rapids and falls and finally joins River Mandakini at Tilwara.

(3) Alaknanda : It originates from Alkapuri glacier at the base of Balkun peak, 6067 masl. It receives on its way R. Dhauliganga, Nandakini, Pindar and Mandakini rivers at Vishnuprayag, Nandprayag, Karnaprayag and Rudraprayag respectively. A number of small streams (*gaderas*) also fall into it. It is a temperamental river, surging through rocky gorges, with rapids and falls.

(4) Badiyar River : It originates at Dandakhil at 1640 masl., lat. N. 30° 20' and long. E. 78° 48' 30", with many rapids, cascades and falls and a long run, falling into Alaknanda near Kotali village.

(5) Jalkur stream at Lambgaun : This stream originates at an elevation of 3000 masl, N. lat. 30° 40' and E. Long. 78° 35'. It joins Bhagirathi on its left bank.

(6) Bhagirathi from Bhadiyana to Tehri : River Bhagirathi originates from an icy glacier at Gomukh, at an altitude of 4071 masl. River Bhilangana meets Bhagirathi on left bank, opposite Tehri town at N. lat 3° 23' 20" and E. long. 78° 3' 1" at 750 masl. A number of tributaries join the Bhagirathi, draining the entire area. Bhagirathi flows first southwest and then southeast after Tehri town and finally joins R. Alaknanda at Deoprayag to form the Ganga onwards to Rishikesh and Haridwar.

(7) Balganga stream at Ghansali (or Bolkhila) : It originates from Kukhli Dhar at an elevation of 4600 masl, N. lat. 30° 45' and E long. 78° 45'. A number of small tributaries fall into it and it finally falls into river Bhilangana at Ghonti.

(8) River Bhilangana from Tehri to Ghansali : This main river of Bhilangana valley originates from Khatting glacier in Tehri, N. lat. 30° 4' 30" and E. long. 79° 1' 3" and takes a westerly course. It has a number of tributaries, with many rapids, falls and deep rock gorges through Bhilang Patti for 80 Kms. It eventually joins R. Bhagirathi on the left bank opposite Tehri town in N. lat. 3° 23' 20" and E. long. 78° 3' 0" at 750 masl.

(9) Streams Bhardari, Dundair, Nailchauri, Aglar, Gular, Nagani and Jalkur : The first four have steep gradients and high velocity. The stream-bottoms are rocky with big boulders and stones. The fish fauna of the last two low-velocity streams is richer than that of the remaining streams.

Altitudinal Differences in Fish Communities

The diversity of the fish community increases from upstream to downstream (Shelford 1911, Thomson & Hunt 1930, Lorimore et al. 1952, Lorimore & Smith 1963, Sheldon 1968). The most likely reason for increase in the number of species downstream is increase in depth of water in areas in lower altitudes and plains (Sheldon 1968). Gorman & Kerr (1978) concluded that changes in fish communities along a stream-order gradient are due to changes in habitat characteristics, and may be reflected by addition to the number of species, replacement of species or by replacement of relative abundance of some species with a set of another species.

The altitudinal distribution of fishes in the streams of Garhwal is correlated with a number of factors : (i) changes in hydroclimate ; (ii) changes in physical factors affected by altitude and gradients ; (iii) changes in stream habitat (ecology) influenced by stream-size, gradient and depth; (iv) changes in biotic conditions such as food, vegetation and other community relations.

Factors of Hydroclimate

Temperature appears to be the most important of the physico-chemical factors in the altitudinal distribution of fishes. *Tor putitora, T.tor* and *Labeo dero* appear to prefer streams maintaining temperatures from 19.5°C - 25°.5 C. *Barilius bendelisis, B. vagra* and *Nemacheilus botia* have a wide range of temperature tolerance from 10°C - 35°C. The temperature tolerance of *Schizothorax richardsoni* ranges from 5°C to 24°.5 C in Garhwal. *Glyptothorax cavia* and *G. garhwali* are found in streams with a temperature range between 5°C - 24.5° C. G. *brevipinnis alaknandi, G. pectinopterus* and *Pseud echeneis sulcatus* are found in temperatures from 8°C - 27°C. *Nemacheilus beavani, N. montanus* and *N. rupecola* are found in water with temperature range of 8°C - 27°C.

The spawning grounds of *Schizothorax richardsoni, Crossocheilus latius, Garra gotyla gotyla* and *Clupisoma montana* have been located not in the main river bed but in the tributaries (shallow) with temperatures between 18.5° C - 21.5° C.

The waters of hillstreams are well-oxygenated. Turbidity, pH, concentration of Carbon dioxide and other gases, and silt are probably of less importance as limiting factors (except in R. Bhagirathi at Tehri where, due to construction of the dam, there is a continuous addition of silt to the water, affecting fishes adversely). However, these latter factors could be of great importance in the determination of absolute and relative abundance of certain fish species.

The fishes are well adapted to the pH in the Garhwal hill streams, where pH is neutral or slightly alkaline.However pH may affect fish food organisms. Usually neutral or alkaline waters do support higher fish population than acidic ones (Troutman 1957).

Stream Gradient

Gradient has a great influence on the distribution of fishes and other aquatic fauna and flora. *Schizothorax richardsoni, Crossocheilus latius* and *Garra gotyla gotyla* are well adapted to steep gradient in the streams and rivers of Garhwal at high altitudes. These fish can live in waters with a velocity of 3 to 5 m/sec. *Glyptothorax* (all species), *Pseudecheneis sulcatus, Tor chillinoides, Nemacheilus beavani, N. montanus, N. rupecola* and *N. botia* thrive in velocities ranging from 3 - 6 m/sec. and rocky bottoms. In Alaknanda the fishes *Mastacembelus armatus, Barilius vagra* and *Nemacheilus botia* are not found above 1311 m (Pipalkote). They inhabit moderate-flow streams (Nagani, Gular and Aglar) where the gradient is not steep and there are only a few rapids and falls. Big boulders, rapid waterfalls and turbulent water-flow are not found in these streams.

The Swim Bladder

The fishes of these hill streams are necessarily bottom-living, because the specific gravity is increased by the reduction of swim bladder. Due to this habit the fish do not make vertical movements and, therefore, do not need a hydrostatic organ. A balloon-like swim bladder would be of great disadvantage. The fishes live mostly under stones or at the bottom.

Bridge & Haddon (1893) stated: "The causes that have led to the degeneracy of the gas-bladder in so many forms are in many instances not difficult to trace, and as in so many physoclist Teleostomi, the assumption of a purely ground habit of life is probably the most important one. Not a few genera of Siluridae abnormales inhabit the comparatively shallow waters of rapidly flowing mountain streams and torrents, often living at a considerable altitude and in general habit are not unlike our common English loaches. Many are provided with an adhesive apparatus on the ventral surface of the body between the pectoral fins for attachment to stones, so that they may be enabled to withstand the force of mountain torrents. Such fishes, when not in motion by the exercise of their fins, probably rest upon or attach themselves to the river bottom, and the utter uselessness and probable harmfulness of an air bladder as a hydrostatic organ under such conditions is no doubt the cause of its degenerate and rudimentary condition in such Silurids as *Sisor, Pseudecheneis, Glyptosternum, Euglyptosternum, Exostoma, Amblyceps*, etc.

Hora (1922) stated that in *Garra* the reduction in the swim bladder is directly proportional to the strength of the current of the stream in which the fish lives.

In all hill stream fishes under study here, such as *Botia dayi, Nemacheilus beavani, N. montanus, N. rupecola, Glyptothorax cavia, G. garhwali, G. pectinopterus, G. brevipinnis alaknandi* and *Pseudecheneis sulcatus*, etc. the swim bladder is much reduced and is divided into two lateral chambers which are more-or-less connected with each other by a short, narrow transverse tube. Moreover the bladder is wholly or partially encapsulated by a bony case in all these fishes.

Due to turbulent and shooting floor of the water of these hill streams, the specific gravity of these fishes is increased by the reduction of swim bladder and abdominal cavity. These modifications are acquired in response to the necessity of remaining as close to the substratum as possible.

REFERENCES

Das, S.M. 1992. Ecological study of high altitude lakes of Kumaun, India and development of applied environmental research. *Rec. Adv. Fish. Ecol. Limn. Eco-conserv., II* : 107-188.

Gorman, O.T. & Kerr, J.K. 1978. Habitat structure and stream fish communities. *Ecology, 59*(3) : 507 - 515.

Kramer, D.L. et al, 1978. The fishes and the aquatic environment of the Central Amazon Basin, with particular reference to respiratory patterns. *Canad. J. Zool., 56* : 717 - 729.

Mendelson, J. 1975. Feeding relationships among species of *Notropis* (pisces, cyprinidae) in a Wisconsin stream. *Ecol. Monogr., 45* : 199 - 230.

Nath, Surendra 1992. Studies on the bioecology of fishes of Jammu Province (Jammu and Kashmir State) India, Part I : Habitat Ecology and Body shape. *Rec. Adv. Fish. Ecol. Limn. Eco-conserv., II* : 87 - 106.

Zaret, T.M. & Rand, A.S. 1971. Competition in tropical stream fishes : support for the competitive exclusion principle. *Ecology, 52* : 336 - 342.

Rec. Adv. Fish. Ecol. Limn. Eco-conserv.
III : 47-48

6

On a Simple Method for Clearing and Exposing Micro Mone Structures for Tracing their Origin and Development

D.K.MASSEY *

INTRODUCTION

In small vertebrates (particularly fishes) the process of cleaning and clearing muscles to expose fine bone structures *in situ* is a difficult, strenuous and time consuming operation.

Keeping this difficulty in mind the author, during his research project, experimented upon a new and easier method of exposing the Weberian apparatus of Teleosts.

In order to clear the Weberian apparatus *in situ* the author devised his own method of getting a fine jet of super-heated steam as follows :

A heat resistant latex tubing was securely fixed on the escape nozzle of a domestic pressure cooker. A hard glass micro-pipette was attached to the other end of the latex tube. The pressure-cooker containing water was put on a stove to heat. The steam issuing out of the glass pipette had the desired temperature and pressure. The superheated jet of steam was carefully trained on the flesh and other unwanted tissue of the fish at the site. The partially cooked flesh falls away easily without disturbing parts under study. After the removal (or study) of the ossicles the whole area was subjected to further steam treatment, which exposed the fine bone structure upto the otic capsule. All this was done under a stereoscopic binnocular microscope so that each articulation could be studied. This method proved to be excellent, for it revealed precisely the relationship of ossicles and skeleton, sinus impar and its canal where it is situated and the entire otic capsule.

A word of caution is required to be said here. Super heated steam is dangerous to handle, more so because it is under pressure. Careless handling can cause grievous burns. The glass tube and latex tube is very hot to touch and should have a clamp attached for handling. The author suggests that a beaker of water be kept nearby where the pipette can be dipped when not in use. The steam will go on bubbling without any danger. Manipulation becomes easy after a little practice.

For the study of minute bony structures and pin pointing their origin from hatching to fry to fingerling stages, the author found that bulk staining by Alizarin-S is very helpful. Later, the specimens were subjected to clearing by steam, using the above method.

---

* Ph.D.Thesis : "Studies on comparative functional anatomy of Weberian apparatus and gas bladder in some Teleost fishes"- Kanpur University, 1981.

1. Entire specimens were fixed in 10% formalin for 24 hours.
2. Partial dissection was done over the site.
3. The dissected specimens were washed with water and transferred to 1% KOH solution for a couple of days or more according to size of fish.
4. The material was then transferred to the following mixture

   Alizarin red .... 0.1 gm
   KOH .... 10 gm
   Distilled water ....1 litre

The material was kept in the solution long enough to be stained to the desired depth of colour.

5. The material was than washed with distilled water and stored in 70% Alcohol.

Rec. Adv. Fish Ecol. Limn. Eco-conserv.
III : 49-61

7

On The Systematics And Bioecology of the Mahaseer Fishes of India, with a Discussion of The Problem of Their Rapid Decline in Western and Central Himalaya

S. M. DAS

ABSTRACT

A recent research survey on Mahaseers in Kumaun-Garhwal waters (CSIR Mahaseer Project of the author, 1975-1980) has revealed the presence of only two species of *Tor* mahaseers (*Tor tor* and *T. putitora*) in the said waters. In a U.G.C. Project survey of Kashmir waters (Das 1968, Das et al 1970) the present author showed the rare occurrence of only one Mahaseer species (*T. putitora*) in Kashmir and possibly none exist today.

Sen and Jayaram (1982) have recognized only 6 valid species and 3 sub-species of *Tor* Mahaseers in India today. Many others (*T.chilinoides, T. hexastichus, T. dukai* and *T. neilli*) have been ruled out and placed under the genera *Barbus* and *Puntius*. The 6 valid species are: *T. tor* (Ham.), *T. putitora* (Ham.), *T. mussullah* (Sykes), *T. khudree* (Sykee), *T. mosal* (Ham.) and *Tor progenius* (McCl.). The 3 subspecies are : *T. khudree malabaricus* (Jerdon), *T. khudree longispinis* (Gunther) and *T. mosal mahanadicus* (David).

The present paper gives a concise account of the Mahaseer systematics, distribution, bio-ecological parameters (problems of food and feeding, reproduction, regional hazards), introduction into lentic water, causes of decline, induced breeding in Bhimtal and steps for rehabilitation of the Mahaseer fish in India. The present contribution is the first composite account on the Mahaseer fishes of India today and will prove of immense help in future researches on the Mahaseer fishes of India.

INTRODUCTION

The Mahaseers (*Tor* Gray) were abundant in India in the thirties but are gradually becoming more and more scarce in the eightees. Despite the researches of Hora (1939, '40, '41, '42, '43), Menon (1974), Misra (1962), Das (1957, '60, '66, '67, '69, '71, '73, '74, '79, '82, '83) and Das et al (1956, '63, '65,'68, '69, '70, '71, '79, '80, '84), and the latest ZSI publication (Sen and Jayaram 1982) , our knowledge of Indian Mahaseers is incomplete. The systematics, distribution, ecological parameters, problems of food and feeding, migration in the Himalayas, breeding niches and periods, problem of Mahaseer decline in the Country, as well as the problem of induced breeding by hormone injection remain in a preliminary stage; all these parameters have now been brought together in the present paper.

The famous game and fine food fish *Mahaseer* occurs in many rivers in India, yet they are rapidly declining. Out of the six valid species of *Tor* Mahaseers only two species (viz. *Tor tor* and *T. putitora*) occur in the riverine systems of North-Western, Northern and Central Himalayas; while 2 species (*Tor mosal* and *T. progenius*) occur in Eastern Himalayas - upto N. Burma. The remaining two species (*Tor mussullah* and *T. khudree*) occur in the Central and South-Indian (Peninsular) waters,*T. khudree* extending upto Sri Lanka.

The usual catch in Kumaun waters consisted of 5 to 8 Kg. mahaseer specimens from the lakes Bhimtal, Naukuchiatal, Sattal and Khurpatal (none in Nainital); and only 2 to 5 Kg. Mahaseer (*Tor tor* and *T. putitora*) from Kali, Gori, Kosi and Ramganga rivers (original observations). Only 40 years back record catches of 22.5 Kg. and 27.0 Kg. from Malwatal (Corbett 1935, 1945 in observations sent to B.S. Raj) and 66 lbs. (Smith 1946, from Manasbal, Kashmir, Personal Communication) are on record. Recently only one record of a giant Mahaseer (*Tor putitora*) from Bhimtal is available (Pathani 1980).

Little work has been done on the bioecology of Mahaseer fish species. The present paper gives a more or less complete picture of the systematics, ecology and biology of the Mahaseers of the Western and Central Himalayas (*T. tor* and *T. putitora*), worked out in the C.S.I.R. project granted to the author (1975-1980, 1977-1982), along with some data on biology from workers in Peninsular India, to form a succint picture of the Mahaseer fishes of India.

The problem of decline of Mahaseers, both in size and population, has been investigated and the reasons for the decline elucidated. Important remedial measures, as worked out for Kumaun Himalayas (which are applicable to the entire Himalayan region), have been suggested in the present paper, as also to the C.S.I.R.and D.S.T. authorities.The author also extends his findings to Garhwal Himalaya, in a project of C.S.T., U.P.

It may be significantly pointed out that all Himalayan Mahaseers (except the land-locked sparse lake population) migrate annually to their spawning grounds for breeding in high altitude streams, which are now being desecrated. In Kumaun region the fish migrate to the upper reaches of Kali, Gori, Kosi and tributaries of Ramganga rivers. In Garhwal region, the Mahaseers of Alaknanda, Mandakini, Bhagirathi and Bhilangana (the latter having muddy and silt-bearing water) migrate into clear tributaries (Hastar, Bhardari, Badyar, Nagami, Balganga, Aglar, Jalkur, etc.) where ripe males and gravid females are found during May to September.

Much work on Mahaseers remains to be done, especially the continuation of the successful induced breeding done in Kumaun region (Pathani and Das 1979) to other parts of the Himalayas. The remedial measures for Mahaseer decline must be tackled conjointly by I.C.A.R., C.S.I.R., C.I.F.R.I., C.S.T. and University teams, to re-establish the size, population for food and game attraction for foreign tourists. Attempts may also be made to export Mahaseer juveniles to foreign Countries for culture and propagation, even as the exotic mirror carp, grass carp, trout and the minnow have been introduced into India, since the fertilised eggs of mahaseers are capable of being flown without water in special boxes (Kulkarni and Ogala 1979).

Another feasible aspect of Mahaseer development is its introduction into more lentic waters in the hills and plains, as introduced into Kumaun lakes by Ramsay 1856, the first person to introduce the fish fry and fingerlings into Nainital and Bhimtal lakes, the seed being procured from the winding Kosi river. Later, Naukuchia Tal and Sattal were also seeded by Ramsay from Gaula river. The fish thrived in Nainital, Sattal and Naukchiatal lakes, but failed to thrive in Bhimtal. More seed was introduced into Bhimtal in 1878 and Mahaseers of 18 lbs. were reported in 1988 from Bhimtal as well by Walker (1988). This was certainly due to the perseverence and perspicacity of Sir Henry Ramsay, Chief Commissioner of Kumaun from 1856-84. The best Mahaseer fishery in Kumaun today is in lake Bhimtal, where record catches of 22.5 Kg. (45 lbs.) in1955 and a recent catch of 18 Kg. (36 lbs) have been obtained. Lake Nainital has not only lost all its Mahaseers due to pollution, but there is also found a high annual mortality of other fish in the lake. Therefore, before seeding fresh waters with Mahaseer, the physico-chemical parameters of the water must be tested, lest the good seed is sown on barren ground.

I. Systematics and Distribution of Mahaseers

A. Valid species

(1) *Tor tor* (Ham.) : Jammu Province, Himachal, Punjab, Haryana, U.P., Bihar, W. Bengal, Assam, Pakistan, Bangladesh, Burna and China, Inhabits mountain streams, rivers and some lakes and reservoirs of India, both in North India and Peninsular India (not Kashmir). Ganga (Upper) and Narbada, with their tributaries, are its main habitat. It grows up to 1.52 M. and attains a weight of 25 Kg. (*Tor Mahaseer; Turiya* and *Mahashol* are common names).

(2) *Tor putitora* (Ham.) : Jammu & Kashmir, Himachal, Punjab, Haryana, U.P., Bihar, N. Bengal, Assam, Nepal, E. Himalaya, Pakistan, Bangladesh. Inhabits rivers and lakes of Kumaun (except Nainital). It grows upto 2.74 M. and attains a weight of 30 Kg. (*Putitor Mahaseer, Mahashol* and *Jungia* are common names).

(3) *Tor mussullah* (Sykes) : Peninsular waters, chiefly S.Indian rivers Godavari, Krishna and Cauvery. It grows upto 1.22 M and attains a weight of 26 Kg. (*Bhavani Mahaseer, Masund* and *Massulla Mahaseer* are common names).

(4) *Tor khudree* (Sykes) : Jammu Province, Orissa, Peninsular waters e.g. rivers Bhima, Krishna, Koyana, Cauvery, as also Mahanadi Godavari and their reservoirs and tributaries and Sri Lanka. Size upto 1.45 M and weight upto. 22.7 Kg. (*Khudri Mahaseer, Deccan Mahaseer, Yellow Mahaseer, Peela Mahaseer* and Irra-Arrayam (Telegu) are common names).

(5) *Tor mosal* (Ham.) : Burma, Jammu Province, E. Himalaya, Assam, Mahanadi and tributaries; mainly inhabitant of rivers. Size 1.0M and weight 11 Kg. (*Mosal* or *Copper Mahaseer* is the common name).

(6) *Tor progenius* (McCl.) : N.E. Himalaya, Assam, Manipur, Naga hills; mainly inhabitant of rivers. Size 0.75 M and weight 10 Kg. (*Junga* and *Jungapithia* (Assamese) are common names)..

B. Valid subspecies:

(1) *Malabar carp* : *Tor khudree malabaricus* (Jerdon) : Size 1' 3", Weight 2 Kg.
(2) *Ceylon khudree* : *Tor khudree longispinis* (Gunther).
(3) *Mahanadi mosal* : *Tor mosal mahanadicus* (David)

C. Problem species (Uncertain)

(1) *Tor hexastichus* (Manon) = *Barbus hexastichus* (McCl.) (3 feet)
(2) *Tor dukai* (Misra) = *Barbus dukai* Day (2 feet)
(3) *Tor neilli* (Misra) = *Barbus neilli* Day (2 feet)
(4) *Tor chilinoides* (Misra) = *Barbus* (*Puntius*) *chilinoides* McCl. (10")

A Key for the identification of *Tor* species (After : Sen & Jayaram 1982)

1. Head length more than 1/5 of total length ... 2
   — Head length less than 1/5 of total length ... 3
2. Eye diameter 3.2—4.2 in head length; L. 1. scales 23-26, head length equalling depth of body - *T. mosal* (Ham.)
   -Eye diameter 4.5-7 in head length, L.1. scales 25-27, head length *more* than depth of body. - *T.mosal mahanadicus* (David)
3. Head length *more* than depth of body; dorsal spine smooth and strongly developed; snout not tuberculated in breeding season in male; L.1. scales 25 - 28; length of head 1/4- 1/4.5 of total length; lips thickened and protruding. ... *T. putitora* (Ham.)
   —Head length *more* than depth of body; dorsal spine may or may not be strongly developed; snout may or may not be tuberculated ... 4
4. Head length equalling depth of body; L.1 scales 25 - 27. ... 8
   —Head length less than depth of body ... 5
5. Eye diameter 1/3 - 1/4 in head length; snout not tuberculated; L.1. scales 22-27. *T.tor* (Ham.)
   —Eye diameter more than 1/4 of head length; snout may or may not be tuberculated; Head length 1/4 to 1/6 of total length ... 6
6. L.1. scales 26- 27, predorsal scales 10; eye diameter about 1/6 of head length; snout tuberculated. *T. mussullah* (Sykes)
   —L.1. scales 23-24, predorsal scales 10 ... 7
7. Head length 1/4.5 to 1/4.7 of total length; predorsal scales 8-9; L.1. scales 23-24; males smaller than females; lips with labial flaps, larger flaps in males ... *T. khudree longispinis* (Gunther)
   —Head length 1/4.7 to 1/5 of total length. rostral barbels short, not extending to eye; L.1. scales 24; eye diameter 1/5 of head length; upper jaw longer; lips thickened *T.khudree malabaricus* (Jerdon)
8. A black streak behind gill opening; tip of snout fleshy and produced into a semicircular flap; L.1. scales 26; dorsal spine less developed; long compressed head ... *T. progenius (McCl.)*

—No black streak behind gill opening; tip of snout not fleshy, tuberculated; L.1. scales 25-27; eye diameter 1/5.5 to 1/7 of head length ... *T. khudree* (Sykes)

Table I. ***Maximum Size and Weight Recorded for the Five Valid Species of Mahaseers***

T. t or	*T. putitora*	*T. khudree*	*T. mussullah*	*T. mosal*
1.5 m. (Misra1962) *5 feet* (Jhingran (1975)	2.75 m. (Misra 1962) *9 feet* (Jhingran 1975)	1.5 m. (Hora 1943) *4' 9"* (Misra 1962)	1.2 m. (Misra 1962) *4 feet* (Jhingran 1975)	1.5 m 55-75 lbs. *5 feet* (Mac Donald 1943)
57 lbs. wt. (Corbett 1935)	2.75 m.	50 lbs. wt. (Hora 1943)	1.2 m.	5 feet

II. Bioecology of Mahaseers

(1) Food and feeding :

Tor putitora (putitora Mahaseer) of Kumaun is a herbiomnivore (Das and Pathani 1978-a) and *Tor tor* (Tor Mahaseer) an omnivore, the latter feeding on aquatic plants as also earthworms, crabs, water beetles, molluscs, aerial and aquatic insects, shrimps, snails etc, and the former mostly on macrophytes of the lakes, along with some of the animals mentioned for *T.tor*. As a consequence the ratio between the intestinal bulb and intestine (the bile duct opening more anteriorly into the bulb) is larger in *T. putitora* than in *T. tor*. (*The juveniles are more insectivorous than adults*, feeding on plankton diatoms, algae, water fleas, aquatic insects, rotifers and protozoans.

Both Kumaun Mahaseers are *clean feeders*, like the snow trout, *Schizothorax*, of Kashmir. *T. khudree* and *T. mosal* have been recorded to ingest bottom organic matter (BOM), detritus and even sewage and dung, and may be termed *foul feeders* (Sen and Jayaram 1982). Thus it is expected that these two Peninsular Mahaseers (*Tor* species) would be able to compete successfully with the bottom foul feeder *Cyprinus carpio*, that has ousted the clean feeders *Schizothorax* and *Schizopyge* in Kashmir. In addition to high column- and surface-feeding observed in Kumaun lakes, all Mahaseers are usually benthophagous.

Tor tor : The average annual plant food/animal food ratio (results of two years' data from Bhimtal, Das and Pathani 1978) is about 1: 1 , with slight monthly variations. The food consisted of the following taxa, usually in order of abundance : Macrovegetation (aquatic), insects, crustaceans, algae, detritus and sand, diatoms, rotifers, fish scales, protozoans, molluscs, nematodes and bryozoans. An annual average range of food items is as follows :

Macrovegetation	-	20 to 30%	Protozoans	-	2 to 5%
Algae	-	5 to 20%	Molluscs	-	1 to 2%
Insects	-	10 to 30%	Bryozoans	-	1 to 2%
Crustaceans	-	10 to 30%	Fish scale remains	-	2 to 8%
Diatoms	-	2 to 5%	Nematodes	-	2 to 4%
Rotifers	-	2 to 5%	Sand & detritus	-	5 to 12%

The large Mahaseers do not prey or feed on young specimens of *Tor* or other fishes as mentioned by Raj (1945), as a 3 - year study by the present author has shown (CSIR Report, 1979). *Tor tor* is thus a typical omnivore.

Tor putitora : Das and Pathani (1978) reported upon the food and feeding habits of the Putitor Mahaseer from Bhimtal lake, and recorded the following sequential order of food intake in adults :-

Table II : Fluctuations in % Plant - Animal Food in the Warm months of high Feeding in T. putitora (Das and Pathani 1978)

Month		*Plants %*	*Animal %*	*Others %*	*Plant/Animal ratio*
April	-	66.0	30.00	4.00	2.20
May	-	69.9	26.08	4.02	2.69
June	-	53.0	41.00	6.00	1.30
July	-	53.7	40.50	5.80	1.32
August	-	58.0	40.00	2.00	1.45
September	-	49.0	47.00	4.00	1.04

Thus *T. putitora* may be called a herbiomnivore, as there is usually a higher intake of plant food than animal food, which latter is also considerable. The decline in plant food/ animal food ratio from April to September indicates the gradual diminution in intake of macrophytes from summer to autumn.

Table III : Comparative RLG, K - Factor and GSI for Mahaseers and Some Freshwater Food Fishes of India. (Das and Pathani 1978; Das and Srivastava 1979)

Fish species	*RLG*	*Highest K-factor*		*GSI*
Herbivores :				
Cirrhinus mrigala	23.92	2.19	East U.P. Lentic Waters.	Not reported
Catla catla	7.27	3.12		-do-
Labeo rohita	11.55	2.43		-do-
Labeo calbasu	13.20	2.57		-do-
Omnivores :				
Tor tor	2.06	1.80	Kumaun Lakes.	3.65
Tor putitora	2.16	1.50		3.85
Puntius sarana	3.30	3.24	East U.P. rivers.	Not reported
Barbus hexastichus	2.30	2.63		-do-
Carnivores :				
Wallago attu	0.63	1.80	East U.P. rivers.	-do-
Channa punctatus	0.44	2.63		-do-
Notopterus chitala	0.40	0.95		-do-
Bagarius bagarius	0.80	2.20		-do-

Feeding Intensity :

The feeding intensity of *T. tor* varied from 3.33 (March) to 13.33 (August), being 5.0 (June), reflecting higher rate of feeding and higher available food in August during the monsoon rains. *T. tor* showed a characteristic inverse correlation between algae and aquatic macrophytes in the gut. The highest percentage of Diatoms was in December - January, when the other food items are scarce in the lakes and the fish feeds at bottom or in the inshore littoral region. Both insects (*Chironomus* and *Chaoborus*) and crustaceans (*Ceriodaphnia* and *Diaptomus*) were preferred animal food; thus the fish is certainly a complete omnivore, when compared to the herbi - omnivory of *T. putitora* (Das and Pathani 1978, Pathani 1982). These results of Kumaun *T. tor* are contrary to those recorded by Desai (1970) for the Narmada *Tor* Mahaseer, where he found it to be a complete herbivore with macro - vegetation as the chief diet. Then again Desai reported high feeding in winter and low in summer, which is just the reverse of Kumaun *T. tor*, the latter having high feeding in summer and low in winter (Similar to Kashmir *Schizothorax, Subla* and Das 1970).

The *Tor* Mahaseer being an intermittant breeder, the fish feeds actively during pre- and post-spawning periods, when food intake is high and the fish may be called an intermittant feeder, which is reflected in the scales, otolith, operculum and vertebrae as growth rings.

***Qualitative food analysis of Tor tor of Bhimtal* (Pathani 1982)**

Macro - vegetation	:	*Polygonum, Myriophyllum, Ceratophyllum, Vallisneria, Hydrilla.*
Algae & diatoms	:	*Ulothrix, Hormidium, Binuclearia, Coelastrum, Pediastrum, Closterium, Pithophore, Ceratium, Synedra, Nitzchia, Cymbella, Diatoma.*
Crustacea	:	*Daphnia, Ceriodaphina, Graptolebris, Bosmina, Cypris, Eucypris, Diaptomus, Cyclops.*
Insecta	:	*Chironomus, Chaoborus, Pethomis, Berosus* larvae; and adults of Coleoptera, Diptera and Hemiptera as well as aerial insects.
Rotifera	:	*Monostyla, Aphelenchoides, Colurella.*
Protozoa	:	*Euglena, Bursaria, Paramoecium, Stylonichia.*
Bryozoa	:	Statoblasts of *Plumatella.*
Mollusca	:	*Lymnaea, Indoplanorbis.*

Food and Feeding of Kumaun Mahaseer Fingerlings

Food studies of fish juveniles have an important role in fish ecology and fisheries, since the growth and production of adult fish depends upon good and sufficient food for the juveniles and young fish. Some recent contributions on the food of fish juveniles are by Alikunhi (1952, mirror carp), Jhingran (1968, *Catla*), Jan and Das (1970, eight Kashmir fishes), Sehgal (1971) , Khan and Jhingran (1975, *Labeo*) and Das and Pathani (1978, *Putitora* Mahaseer). All these workers have shown that fish juveniles have different food and feeding habits from the adults.

Fingerlings of *T. tor* and *T. putitora* feed and move about in shoals of their own age-groups upto about one year of age in lake Bhimtal near the perennial Mallital streamlet inlet. They are more zoophagous than the adults. The maximum food in their guts consisted of adults, and larval insects, followed by Crustaceans, macrovegetation, diatoms, rotifers, protozoans, sand and organic debris. This zoophagous habit also affects the RLG ratio, which is far less than in adult Mahaseers, ranging from 0.70 to 1.00 in both species (Pathani and Joshi 1980). The ratio of plants to animals (Plant food/animal food ratio) is also low, as seen in lakc Bhimtal, Kumaun.

The aquatic vegetation in gut contents consisted of soft parts of *Myriophyllum*,

Ceratophyllum, Hydrilla and *Vallisneria;* the algae were mostly *Spirogyra, Binuclearia* and *Siderocelis;* the diatoms *Synedra* and*Nitzschia*; the Crustaceans *Daphnia, Ceriodaphnia, Cyclops,Cypris,Condona* and *Nauplius* larvae; the insects were represented by Coleopteran juveniles and adults, *Chironomus (Tendipes), Chaoborus* and other Dipteran larvae; the only rotifer recovered was *Brachionus* (an oligotrophic form); and the protozoans *Centrophrix* and some remains of Ciliates (Pathani and Joshi 1980).

Sehgal reported on the food of fry and fingerlings of Mahaseer (Sp. not known), belonging to size range 25mm to 130 mm to consist of :-

Bacillariophaceae	-	46.98%	Rotifers	-	1.42%
Decaying DOM	-	22.28%	Bluegreen Algae	-	1.38%
Green Algae	-	13.33%	Protozoans	-	1.38%
Insects	-	10.21%	Fish remains	-	0.41%

His results are somewhat problematical as he does not state whether these are annual averages, since there is much fluctuation in food intake from month to month. Besides, as observed by the present author, fry and fingerlings do not eat other fish; and Sehgal's "fish" in gut would possibly indicate the scales present in the gut, often ingested during bottom feeding along with DOM. Again Sehgal's high value of DOM (22.28%) may really represent digested food rather than ingested DOM, since Mahaseer juveniles are clean feeders like the adults, although bottom feeding may cause incidental ingestion of some sand and organic debris.

The food contrast between juvenile and adult *Tor tor* of Narbada (Karamchandani 1967) is as follows :-

Food items	*Adult*	*Juveniles*
Macrovegetation	41.7%	4.7%
Algae	12.3%	2.8%
Molluscs	10.5%	21.9%
Insects	8.0%	53.8%

Table IV : *The RLG and average plant food/animal food ratio of the fingerlings of T. tor and T. putitora **

	Total length in mm.	*Number Examined.*	*Length of gut in mm.*	*RLG Range*	*Plant/Animal ratio*
Tor tor :	31.0-40.0	20	29.0-38.0	0.70-0.90	0.49
	41.0-50.0	19	38.0-50.0	0.78-1.00	0.64
	51.0-60.0	24	39.0-61.0	0.79-1.00	0.67
Tor - putitora :	31.0-40.0	26	21.0-33.0	0.56-0.84	0.41
	41.0-50.0	25	28.0-39.0	0.72-0.90	0.46
	51.0-60.0	31	35.0-55.0	0.73-0.95	0.53

*Data from Pathani and Joshi (1980)

These data for fingerlings of *T. tor* and *T. putitora* show decisively the carnivore type low RLG (0.56 -1.00) and low plant food/animal food ratios.

(2) Breeding Season

It has been confirmed by recent work on Kumaun Mahseers that : (a) Mahaseers do not breed only once (as with *Labeo, Catla, Cirrhinus),* but bread 3 times a year, having four batches of ovarian eggs, the largest at the front end of the ovary and the smallest at the hind end not shed till next year; (b) in cold waters of Kashmir, H.P., Nepal and Kumaun, they spawn intermittently (3 times) from May to September. Thus juveniles of 3 sizes are found during the year. All the ovarian eggs are never deposited at the same time, leaving the ovaries completely vacant, as stated by Khan (1939) and some others, laying all its eggs at one time. Nor does Mahaseer "lay an egg a day for many days just like a fowl" as opined by Thomas (1897).

Tor tor has a longer breeding season than *T. putitora* and shows high spawning migration swimming upstream to head waters and source streams of a river,where they usually spawn just before the summer monsoon rains, and return by November in a spent condition. The lake Mahaseers (like lake trout) may spawn inside the lake or reservoir, in shallow regions, especially near water inlets if available. The spawn is laid after nuptial play (clearly observed in oligotrophic lake Bhimtal) on rocky or stoney substratum in littoral regions of the lake or on rocks and gravel in streams in shallow water or in rock-pools. The male usually entwines around the female (as with other carps) and compresses the abdomen. The adhesive eggs are laid in clusters which adhere to the rocks, and the male sheds milt over them for external fertilisation, while the male vent lies in juxtaposition to the female vent. Although there is no copulation, both roe (eggs) and milt (sperms) are shed at the same time. The theory that many males shed milt over the laid eggs, in the manner of a gardner sprinkling a flower-bed, must be only a conjecture. Some accompanying males over one female may shed milt in excitement but only one male embraces one female fish. It was observed in Kumaun lakes that no spawning takes place on a muddy, loamy or DOM substratum (and lake Nainital is without Mahaseer fishdue to such a substratum (and lake pollution). The fish may also move up small canals or streams (Mallital stream in Bhimtal) that fall into the lake to spawn ; and the fry and fingerlings are usually massed in such spots in the lake.

Tor khudree (Deccan Mahaseer) spawns between June and December; the fish may spawn successfully in confined reservoir waters (as with Kumaun lakes Mahaseers). *T. progeneius* has breeding habits similar to those of *T. tor;* while the breeding behaviour of *T. mosal* has still to be reported upon. It has been observed that younger Mahaseers breed earlier in the season than older ones (Jhingran 1975).

(3) Sex Dimorphism and Secondary Sexual Characters

Macdonald (1943) stated "Sex dimorphism is only possible when ripe females show a fully budged soft abdomen with pinkish vent". But many authors have recorded secondary sexual characters to differentiate the male from the female. Hora and Mukerjee (1936), Hora (1943), Munroe (1955), Jhingran (1975) and Pathani and Daś (1980) have all depicted one character or the other to differentiate the sexes in *Tor tor, T. putitora, T. mussullah* and *T. khudree,* which are given in Table V.

(4) Spawning, eggs and hatching

Spawning (ovulation) usually occurs in high hill streams (or in land-locked waters in currents regions), at water temperature ranging (for *Tor tor)* from 16° to 24°C, depending

Table V : *Secondary Sexual Characters of Male and Female Mahaseers*

Character	*Tor tor*	*T. putitora*	*T. massullah*	*T. khudree*
Snout tubercles	Absent in females, few in males.	None in females, few in males.	Numerous and prominent in males; few in females	Snout with groove; no tubercles in male or female.
Pectoral fin	Equal in size, and colour in both sexes	Bright orange rims in males; light pink in females.	Equal in size and colour in both sexes.	Similar in both sexes.
Dorsal spine	Similar	Thicker and longer in males	Stronger in male.	Similar.
Lip thickness	Thicker and protruded in males.	Lips more protruded in males than females	Lower lip with thickened tip in male.	Hyper-trophy and thicker in males.
Body size	Similar	Females larger than males of the same age.	Similar	Males usually smaller than females.
Labial flaps	Similar, if present	No flaps in both sexes.	Small flaps in males, none in females.	Thicker and longer flaps in males.
Anal fins	Greyish in both sexes	Reddish in male pinkish in female.	Greyish in both sexes.	Greyish, but obliqu-ely placed in females.
Upper jaw	Similar in both sexes.	Similar in both sexes.	Similar in both sexes.	Upper jaw thicker and protruding in males.

upon the nature and location of the water body. After ascending to the spawning ground (spawning migration) *Tor* Mahaseer selects a rock, stones, gravel on sand bars in shallows, and after the usual sex-play, deposits the eggs which adhere to the substratum and do not flow downstream with the current. The spawning starts at the onset of rains in June or snow-melt floods in May, which change the physico-chemical parameters of the water and stimulate production of reproductive hormones. In Kumaun lakes sex play starts at the onset of pre-monsoon showers (first week of June), and first spawning occurs at the end of June or the first week of July (CSIR Report, Das 1978). It may be stated that the Bhimtal observations for ovulation, eggs and hatching time are more in line with the observations of Kulkarni (1979, 1980).

The observation of Jhingran (1975) for *T. putitora* of Punjab, that it spawns three times a year, (i) January-February, (ii), May-June, (iii) July-September, does not hold good for Kumaun Putitor Mahaseer. In the rivers and lakes of Kumaun (lakes Bhimtal, Naukuchiatal,

Sattal, Khurpatal and the rivers Kali, Gori, Kosi, Ramganga), the three spawnings are closely apposed : (i) May-June (ii) July-August (iii) September-October, but not necessarily in these bimonthly periods as usually the next batch of Ovarian ova become ripe within two to three weeks of the ovulation of the last batch. Here also our observations tally with those of Karamchandani (1960) who reported spawning at intervals of 2-3weeks in between two acts; and as soon as the last batch of ova is shed, the ovary becomes again mature and gets ready for the next spawning. 2-3 weeks appears a short time for immature Ovarian ova to mature and also to become ripe for shedding. But it is certain that the first spurt of spawning provides a feed-back mechanism to the pituitary to secrete more hormones (FSH) for fast action on the ovarian follicles, for the next spawning.

T. khudree spawns in S. India at the end of July and August, the fish moving up the rivers at the onset of South-West monsoon at the end of May; and also at the beginning of N.E. monsoon during October or November (Lovatt 1956). It spawns intermittently for 7 months of the year between June and November, while Kulkarni (1954) mentioned only June-August (3 months). *Tor mussullah* and *T. progenius* have similar breeding habits as *Tor tor*. Little is known about the spawning habits of *Tor progenius*. Mahaseers rarely spawn during March, as the water temperature in the hills is less than 17° C, the lowest spawning temperature for *Tor tor* of Kumaun being 20° C, showing temperature to be an important parameter for Mahaseer breeding.

Eggs of *Tor tor*, *T. putitora* and *T. khudree* are heavy, demersal and full of yolk. Egg diameter of *T. khudree* is 3.2 mm and of *T. tor* 2.8 mm (Kulkarni 1980). Desai (1970) reported ova size of *T. tor* as only 1.0-2.2 mm.This must be due to difference between mature and immature batches of ovarian ova. Eggs of *T. putitora* measure 3.0 mm on an average (Pathani and Das 1980) , in which the 3 spawning spurts can be indicated by the 3 ova sizes in ovaries during the spawning period.

Mahaseer eggs take 72-85 hours to hatch out in water after the spawning act, at water temperatures 22°-26° C (Kulkarni 1979,1980). In the lower summer water temperature of Bhimtal (18°-20° C) hatching occurred after 75 to 90 hours of spawning (Pathani and Das 1979). The larvae after hatching have a semiquescent stage lasting 6-7 days for *T. khudree* (Kulkarni and Ogale, 1979), while in the Bhimtal *T. tor* the semi-quescent stage lasted 8-9 days.

(5) Problem of Decline of Mahaseer in Size and Population :

Although the problem of decline of Mahaseer fishes is an all-India one, there has been a more rapid decline in the Western and Central Himalaya. Thus the Mahaseer rod and line sport (one main attraction for the tourists in the past) of baiting the mahaseer has also declined considerably.

Putitor and Tor Mahaseers mature at an age of about 2 years, and live for 10-12 years (personal observation by scale studies of an 18 Kg., 335 mm long specimen). Their total fecundity is from 6000 to 22,000 in *T. putitora,* according to age and size (Das and Pathani 1978); while that for *T .tor* is from 6,000 to 43,000 (Karamchandani 1969). The recent catch of 18 Kg yielded and egg count of 356,500 (Pathani 1980). Therefore, there is no apparent reason that they should decline so rapidly. But a Survey of Kashmir (Das and Subla 1970) Punjab and Kumaun - Garhwal (CSIR Report, Das 1979) has shown diminution in both size and population of the famous Mahaseer. Old records of 22 to 75 lb Mahaseers (Macdonald 1943) are uncommon. In the personal knowledge of the author, one 57 lb. Mahaseer was caught in lake Nainital (Corbett 1935); another 65 lb. one from lake Manasbal in Kashmir (Das 1970); and one from Bhimtal of 37 lbs. (Pathani 1978).

The main reasons for the decline of Mahaseer in India may be summarised as follows:

(i) Capture of breeders, juveniles and poaching during closed breeding season in all the hills;

(ii) Unscientific capture of fishes by building temporary stone dams across hill streams and using fine - mesh net or cloth by village people, whose population has increased 3-4 times than that of the thirties;

(iii) Using of dynamite or hand grenades by many agencies in the higher streams, killing off shoals of large breeder fish for food;

(iv) Poisoning of streams and rivulets by local poisons (extracts of *Derris, Chenopodium, Euphorbia, Artimisia, Croton* etc.) to kill and catch whole schools of Mahaseer and other fishes, small and large;

(v) Increasing large-scale collection (exploitation) of stones, gravel, sand etc. (the spawning inches of Mahaseer) as around Dehra Dun and Garhwal river-beds.

ACKNOWLEDGEMENTS

The author is thankful to the CSIR, Delhi for granting a three and a half year project on "Ecology and Decline of Mahaseers in Kumaun Himalayas". He is grateful to Dr. D.D. Pant, Emeritus Professor (and Ex-Vice Chancellor) Kumaun University for providing research facilities in the University.

REFERENCES

Chaturvedi, S.K. (1976) :*J. Bombay Nat. Hist. Soc., 73* (1): 336-344.
Corbett, Jim (1935) : *"Maneaters of Kumaun"*.
Das, S.M. (1979) : *CSIR Mahaseer Project Report,* CSIR, New Delhi.
Das, S.M. (1979) : Environment Management, U.S.A., *3* (5) : 385.
Das, S.M. & S.S.Pathani (1978) :*Indian J. Anim.Sci., 48* (6) : 461-465.
Das, S.M.& S.S.Pathani (1978) : *Proc. DST. Nat. Seminar on Natural Resources,Dev, & Env* 498-499.
Das, S.M.& A.K. Srivastava (1979) : *J. Inld. Fish. Soc. India, 11* (1) : 6-11.
Das, S.M.& B.A. Subla (1970) : *Kashmir Sci., 7* (1-2) : 24-30.
Das, S.M. & Surendra Nath (1965) :*Ichthyologica, 4* (1-2) : 63-78.
Das, S.M. & Surendra Nath (1971) : *Kashmir Sci., 8* : 1-22.
Das, S.M. & S. Hussain (1970) : *Ibid., 7* (1-2): 85-90.
Das, S.M. & S.K. Moitra (1956) : *Proc. Nat. Acad. Sci., India, 24* (4) : 224-232.
David, A (1953) : *J. Asiat. Soc. Calcutta, 19* : 197-209.
Desai, V.R. (1970) :*J. Inld. Fish. Soc. India, 2* :101 -102.
Desai, V.R. (1972) : *J. Zool. Soc. India, 24* (1) : 47-51.
Desai, V.R. (1973) : *Proc. Indian Nat. Sci. Acad., 19* : 228-248.
Hora, S.L. (1939) : *J. Bombay Nat. Hist. Soc., 41* (2) : 272-285.
Hora, S.L. (1940 a) : *Ibid. 41*(3) : 518-525.
Hora, S.L. (1940 b) :*Ibid., 41* (3) : 784-794.
Hora, S.L. (1941) : *Ibid., 42* (3) : 526-672.
Hora, S.L. (1943) : *Ibid., 44*(1) : 1-8.
Hora, S.L. & K.S. Misra (1941) : *Ibid., 42* : 305-309.
Jan, N.A. & S.M. Das (1970) : *Kashmir Sci., 7* :102-111.
Jhingran, V.G. (1975) : *"Fish and Fisheries of India"* Hindustan Publ. Corp., Delhi.
Karamchandani, S.J. V.R. Desai & M.D. Pisolkar (1967):*Cent. Inld. Fish. Res. Instt., 19* : 1-39.
Khan , H. (1939) : *J. Bombay Nat. Hist. Soc., 40* : 653-656.
Kulkarni C.V. (1971) : *Ibid., 67* : 510-521.
Kulkarni, C.V. (1979) : *Aquaculture, 16* : 367-368.
Kulkarni, C.V. (1980) : *J. Bombay Nat. Hist. Soc., 99* (1) : 70-75.

Kulkarni, C.V. & S.N. Ogale(1979) : *Ibid., 78* (3) : 651-660.
Lal, M.B, & P. Chatterjee (1962) : *J. Zool . Soc. India, 14* (1-2) : 230-243.
Lovatt, J.D. (1956) : *J. Bombay Nat. Hist.* Soc., *54* (1) : 197.
Menon, A.G.K. (1974) : *J. Inld. Fish. Soc India, sp. pub.* (1) : 45-47
Macdonald, S.St. J. (1942-44) : *J. Bombay Nat. Hist. Soc., 42* (2) :*44*(1); *44*(3); *44*(4).
Munro, I.S.R. (1955) : *Marine and Freshwater Fishes of Ceylon. Canberra,* Ext. Aff. Dept.
Misra, K.S. (1962) : *Rec. Indian. Mus., 57* (1-4): 149-156.
Nath, Surendra (1982) : *J. Bombay nat. Hist. Soc., 79* (2) : 430-431.
Nath, Surendra (1986 a) : *Livestock Adv., 11* (12) : 31-36.
Nath, Surendra (1986 b) : *J. Zool. Soc. India, 38* (1-2) : 83-98.
Pathani, S.S. (1981) : *Matsya, 7* : 41-46.
Pathani, S.S. & S.M.Das (1979) : *Sci. Cult., 45* : 209-210.
Pathani, S.S & L.M.Joshi (1980) : *Bioresearch, 4* (1) :43-46.
Raj. B.S. (1945) : *Proc. Nat. Inst. Sci. India, 3* :344-345.
Sehgal, K.L. (1972) : *Silver Jubilee Volume, Centr. Inld. Fish. Res. Instt.,* 125-131.
Sen, T.K. & K.C. Jayaram (1982) :*Rec. Ind. Mus. Occ. Pap., 39* : 1-39.
Sykes, W.H. (1840) : *Ann. Mag. Nat. Hist., 4* : 54-62.
Sykes, W.H. (1844) : *Trans. Zool. Soc. London, 2* : 349-378.
Thomas, H.S. (1897) : *Rod in India,* London.
Tripathi, Y.R. (1978) : *J. Inld. Fish. Soc. India, 9* : 161.
Vasisht, H.S. (1959) : *Res. Bull. Punjab Univ. Sci., 10* : 65-70.
Walker,W. (1888) : *Angling in Kumaun Lakes.*

Rec. Adv. Fish Ecol. Limn. Eco-Conserv.
III : 63-69

8

Studies On The Bioecology of Fishes of Jammu Province (Jammu and Kashmir State) India, Part II : Feeding Ecology

SURENDRA NATH

Fishes consume a great variety of food and there are many modes of feeding. On the basis of the nature of the food taken, these consist of often overlapping categories, viz. herbivores and detritophages, omnivores (consuming small invertebrates) and carnivores (consuming fishes and higher invertebrates). Herbivores, omnivores and carnivores can be found in the same family.

According to the diversity of their food, fishes are classified as (i) euryphagous, with a mixed diet, (ii) stenophagous, with a limited assortment of types of food, and (iii) monophags, consuming only one sort of food, e.g. mollusc feeders, crustacean feeders, etc. According to the feeding conditions some other ecological groups are recognised, e.g. pelagic plankton-feeders, benthos-feeders, etc.

Das and Moitra (1955, 1956, 1958, 1963) studied the food, feeding habits and morphology of the alimentary canal of 24 species of freshwater teleosts of Uttar Pradesh (India). Taking the percentage of ingested foodstuffs as the criterion, they classified these fishes into three primary groups, viz. the herbivores, the omnivores and the carnivores. According to their estimations, a herbivore is one in which more than 80% of the average annual feed is formed by algae and higher aquatic plants. The omnivores feed approximately 50% both on plant and animal food, the plant food ranging from 11.9% to 79.1%, and the animal food from 14.6% to 78.1% in their average annual feed. The carnivores feed more on animal food which forms more than 80% in their average annual feed.

Further, taking feeding affinities of fish to particular water levels (according to their specific foods and the fauna and flora present at those levels) as the criterion for classification, Das and Moitra (1955) also divided fishes into three secondary groups, viz. the surface-feeders, the mid-feeders and the bottom-feeders. The surface-feeders feed mainly on planktonic and surface or sub-surface organisms, consisting of plankton algae, rotifers, crustaceans and their larvae and insects. The mid-feeders or column feeders feed both on surface and bottom food organisms such as algae, aquatic plants, insects, bryozoans, adult crustaceans, fish and fish-scales and mud and sand in addition to the characteristic mid-food organisms. The bottom-feeders feed mainly on the benthonic organisms such as the bottom crustaceans, insects, bryozoans, molluscs, fish and fish scales and large quantities of decomposed aquatic vegetation, mud and sand. According to Das and Moitra (op.cit.) the "surface-feeders are mainly omnivorous, or carnivorous", while the mid-and bottom-feeders are herbivores, omnivores or carnivores ". Apparently competition for food is the greatest in the bottom layers of the water bodies, while there is almost similar competition in the surface and middle layers.

In the ichthyofauna of Jammu Province *Schizothorax richardsonii, Schizopyge progastus, S. longipinnis, S. curvifrons, Labeo dero, L. rohita, L. dyocheilus, L. calbasu, L. bata, L. pangusia, L. microphthalmus, L. boggut, Garra gotyla gotyla, G. lamta, Crossocheilus latius, C. latius punjabensis, Amblypharyngodon mola, Cirrhinus reba* and *C. mrigala* are herbivorous teleosts in whose average annual feed algal and higher plant material comprises more than 70% of the total annual food intake.

The omnivorous teleosts of Jammu are : *Barilius vagra, B. barila, B. bendelisis bendelisis, B. bendelisis chedra, Puntius conchonius, P. chola, P. sarana sarana, P. sophore, P. ticto ticto, P. amphibia, Tor putitora, T. khudree, T. mosal, Salmostoma bacaila, Chela cachius, Esomus danricus, Danio (Danio) devario, D. (Brachydanio) rerio, Rasbora daniconius daniconius, Rasbora rasbora, Colisa fasciatus, Aspidoparia morar, Osteobrama cotio, Eutropiichthys vacha, Mastacembelus pancalus* and *Macrognathus aculeatus.*

The carnivorous teleosts of Jammu are : *Noemacheilus botia botia, N. marmoratus, N. rupecola, N. prashari, N. gracilis, N. kashmirensis, N. vittatus, N. sinuatus, N. montanus, N. striatus, N. kangrae, N. corica* (all entomophagous), *Botia birdi, B. rostrata, B. almorhae, Lepidocephalus guntea, Glyptothorax kashmirensis, G. conirostre, G. cavia, G. pectinopterus, G. telchitta, Glyptosternum reticulatum, Bagarius bagarius, Aorichthys seenghala, Mystus vittatus, M. bleekeri, Ompok bimaculatus, Wallago attu, Clupisoma garua, Amblyceps mangois, Heteropneustes fossilis, Xenentodon cancila, Channa orientalis, C. punctatus, C. striatus, C. marulius, Mastacembelus armatus, Chanda nama, Badis badis, Glossogobius giuris, Notopterus notopterus* and *N. chitala.*

Seasonal fluctuations in the amount of feed is more marked in the herbivores and the carnivores due to difference in available food in the environment. The omnivores show little variation in the feed during the year. This is evidently due to their adaptability to both types of food, and when one kind of food is low in its annual cycle in the environment, the other is consumed in larger amounts by the fishes to have normal feeding. Further, on the basis of feeding affinities to particular water levels where their specific food items abound, the Jammu fishes can be classified into the three secondary groups - the surface-feeders, the mid-feeders and the bottom-feeders.

The surface-feeders feed mainly on planktonic and surface or subsurface organisms consisting of plankton algae, rotifers, crustaceans, insects and their larvae. Fishes belonging to this group are : *Eutropiichthys vacha, Ompok bimaculatus, Glossogobius giuris, Chanda nama* and *Salmostoma bacaila.*

The mid-feeders feed both on surface food organisms and bottom food organisms in addition to the characteristic column organisms. These fishes are neither true surface-feeders nor true bottom feeders, but wanderers between the two zones, although in the main they are confined to the mid regions. Examples of this group are *Labeo rohita, Colisa fasciatus, Mystus vittatus, Notopterus notopterus, Notopterus chitala, Aorichthys seenghala, Wallago attu* and *Xenentodon cancila.*

The bottom-feeders feed mainly on benthonic organisms such as bottom crustaceans, bryozoans, molluscs, large quantities of decomposed aquatic vegetation, mud and sand. Fishes belonging to this group are : *Schizothorax richardsonii, Schizopyge progastus, Labeo calbasu, L. bata, L. dero, L. dyocheilus, L. pangusia, L. microphthalmus, Garra gotyla gotyla, G. lamta, Crossocheilus* sp., *Cirrhinus mrigala, C. reba, Amblypharyn godon mola, Puntius sarana sarana, P. sophore, Osteobrama cotio, Noemacheilus sps., Channa* sps., *Glyptotho-*

rax, Glyptosternum, Bagarius bagarius, Mastacembelus, Clupisoma garua and *Amblyceps mangois.*

The featherbacks —*Notopterus chitala* and *N. notopterus* - feed on insects, crustaceans and small pelagic fish. They are typical carnivorous forms and prefer large rivers rich in aquatic vegetation which have an abundant insect and crustacean fauna. The plant food forms about 15% of the diet while animal food represents 80% of the dietary in the average annual feed.

The Schizothoracine fishes—*Schizothorax richardsonii, Schizopyge progastus, S. longipinnis,* and *S. curvifrons*—feed almost entirely on aquatic vegetation. *Schizothorax richardsonii* creeps over smooth stones, scraping off the green vegetation (algae and moss) covering them; it also ingests a few trichopterid, chironomid and caddis fly larvae attached to the stones, along with lumps of gravel and mud mixed with pulpy vegetable matter.

The *Rasbora* danios are omnivorous teleosts which feed actively on crustaceans, aquatic annelids and soft aquatic plants, the animal component predominating in their average annual feed. The fry ingest large quantities of infusoria, rotifers and microworms.

Barilius sps. are omnivorous teleosts which feed on aquatic annelids, microcrustaceans, rotifers and soft aquatic plants; the animal component predominates in their average annual feed.

The Barbins almost all feed on the bottom or on plants, small organisms and algae.

The Labeos are typically herbivorous species, some of which can scrape off encrusting periphyton from the surface of submerged stones with the horny covering of their lower lip, e.g. *Labeo dero*. Moreover, some species use their lips as sucking organs. *Labeo rohita* feeds on weeds and aquatic vegetation which form about 76% of the average annual feed, the animal component being about 12%. The animal food is represented by bryozoans, rotifers, insect larvae and crustaceans. *Labeo dero* is in the habit of feeding chiefly at night, quietly grazing away at the tender leaves and shoots of aquatic plants. It resorts to plankton and mid feeding during the day. The sand grains often found in its gut contents also suggest a benthophagous habit. *L.calbasu* supplements its plant food with a small quantity of crustaceans and rotifers. In *L. bata* the animal food was almost entirely absent in the dietary, while plant food formed about 80% of the average annual feed, the rest being mainly sand.

The Mahaseers (*Tor* sps.) feed at dusk, primarily on insect larvae, especially the larvae of the Mayflies (*Palingenia* and *Polymytarsis*) and also crustaceans, worms, fish spawn and some plants, being thus omnivorous in their feeding habits; the animal component predominates in their average annual feed and are thus labelled carni - omnivores.

Puntius sarana, P. sophore, P. conchonius, P. chola, P. ticto, P. amphibia are omnivorous in their feeding habits. *P. sarana* feeds on plant food (about 70%), and animal food consisting of bryozoans (2%), insects (7%), crustaceans (3%) and molluscs (12%). In *P. sophore* plant food comprises about 58% and animal food about 30% in the average annual feed. The animal food consists of bryozoans (3%), rotifers (7%), insects and their larvae (6%), crustaceans (13%) and molluscs (1%).

Osteobrama cotio feeds on a variety of food and is omnivorous in its feeding habits. The plant food forms about 32% and animal food about 58% of the average annual feed. *Salmostoma bacaila* is another omnivore in which plant food forms about 12% and animal food 79% in the average annual feed. The animal food mainly consists of insect larvae and crustaceans.

Cirrhinus mrigala and *C. reba* resemble *L rohita* in being true herbivores. They ingest debris and algal encrustations on mud, together with a small quantity of crustaceans.

Crossocheilus latius is a predominantly benthophagous herbivore, which feeds mainly on blue - green filamentous algae and higher aquatic plants. The animal component consists of aquatic insects, insect larvae and microcrustaceans.

Amblypharyngodon mola is a herbivore in which the plant food is 80.9% and animal food about 2.0% in the average annual feed.

Fishes belonging to the Barbin genus *Garra* are benthophagous herbivores. Small groups of this fish (*G. gotyla*) are seen rasping off algal encrustations from the stones in the midst of the stream in River Tawi as well as its tributaries and in Poonch River.

Amongst the Cobitididae, *Lepidocephalus* feeds on insect larvae, worms and crustaceans. The Noemacheilin loaches are predominantly entomophagous and also ingest algae and other food from the muddy river bottoms. *Botia* sps. are typically larvivorous fish, which also ingest a considerable amount of sand, indicating a benthophagous habit. The plant food is of very low percentage and consists mainly of algae and diatoms.

The Bagrid catfishes are typically bottom-dwelling carnivores which hunt for food at night. They are voracious feeders, preying generally on all types of animals. *Aorichthys seenghala* is predatory in its habits and attacks small carps and other small teleosts and prawns. *Mystus vittatus* and *M. bleekeri are* carnivorous teleosts in whose diet plants comprise about 16% of the total food intake, while the animal component is about 82% in the average annual feed. The animal food consists of large quantities of insects, rotifers, crustaceans, molluscs, bryozoans, fish fry and fish scales.

Wallago attu is markedly piscivorous and seeks its prey during the day just under the surface of water. It is not only carnivorous—feeding on fishes, their eggs and fry— but also a voracious and foul feeder which devours carcases. It is regarded as a scavenger of the rivers and feeds both during the day as well as at night. *Ompok bimaculatus* feeds mostly on insects and their larvae, thus being a carnivorous teleost. The plant food forms about 14% and animal food about 86% of the dietary in the average annual feed. The animal component of the diet is mainly composed of insects and their larvae, crustaceans and fish.

Clupisoma is a carnivore which feeds on insects and their larvae, pelagic crustaceans and small fishes. *Eutropiichthys vacha* is omnivorous and subsists on insects, as well as small fishes; the plant food-animal food ratio in its average annual feed is about 27% : 68%. It ingests large amounts of crustaceans (*Cyclops, Gammarus, Daphnia*), rotifers, insects, molluscs and fish.

Amblyceps mangois is a vicious carnivore, in whose diet fishes and aquatic insects and their larvae predominate.

Bagarius bagarius, the largest sisorid, is an active predator which feeds not only on small fishes but also on frogs, fresh water microcrustaceans and other invertebrates. *Glyptosternum reticulatum* feeds on invertebrates, mainly insect larvae — such as those of ephemerids and trichopterids. *Glyptohorax* sps. are benthophagous carnivores which feed on ephemerid and blepharocerid larvae, along with a small amount of aquatic vegetation and inorganic detritus.

Heteropneustes fossilis is a predator which may include large fishes in its diet. Its greatly distensible mouth enables the fish to swallow fishes half its own length in a single piece. This catfish usually feeds on live food of various kinds such as fishes, worms, mussels, etc.

Xenentodon cancila is another predator which darts out at small fish with its long jaws open. The garfish devours quite a large number of small fingerlings.

The murrels (*Channa sps.*) are carnivorous predators which feed on very diverse foodstuffs, particularly on tadpoles. They also feed on worms, insects and fish fry. *C. striatus* is highly carnivorous and takes animal food as mayfly — and dragonfly — larvae, aquatic Coleoptera, *Gammarus*, fish and tadpoles. The plant food is about 5% and animal food about 95%.

Chanda nama is a larvivorous teleost which ingests a large quantity of insect larvae and can, therefore, prove very useful in mosquito control measures. It feeds mainly on animal food (about 98%) which consists of insect larvae, crustaceans and fish scales. The plant food is barely 1% in the average annual feed.

Colisa fasciatus is an omnivore which feeds mainly on large algal filaments and animal food almost 50 : 50, but starved fish could also be fed on animal plankton. The animal food consists of insect larvae, crustaceans and rotifers.

The young of *Badis badis* (leaf fish) are cannibalistic while the adults are voracious feeders which subsist on living animals, usually other small fishes, small worms and crustaceans.

Glossogobius giuris is a benthophagous omnivorous teleost which has a ventrally directed mouth for feeding on small benthonic organisms. The plant food forms about 13% and animal food 88% of the average annual feed. The animal food mainly consists of insects, insect larvae and crustaceans.

The Mastacembelid eels (*Macrognathus* and *Mastacembelus*) are carnivorous bottom feeders and chiefly feed on worms, insect larvae (white worms, blood worms and glass worms) and microcrustaceans (*Cyclops, Daphnia*) swimming through the water. *Mastacembelus armatus* feeds on large crustaceans as well as small fishes.

Modes of Feeding

On the basis of their modes of feeding Lagler et al (1962) classified fishes into five types : predators, grazers, food-strainers, food suckers and parasites.

Predatory fishes feed on macroscopic animals and possess certain adaptations in common, such as the presence of well developed grasping and holding teeth, as in *Aorichthys seenghala, Wallago attu, Bagarius bagarius, Heteropneustes fossilis, Xenentodon cancila*, etc. In all these predators there is a well developed stomach with strongly acidic secretions and the intestine is much shorter than that of the herbivores of a comparable size. Many predators—such as the feather - backs (*Notopterus*), *Wallago attu, Amblyceps mangois, Bagarius bagarius, Heteropneustes fossilis, Xenentodon cancila*, etc. - actively hunt their prey, whereas others, such as the bagrid catfishes, *Clupisoma*, the murrels, etc. often lie in wait till an animal passes by and then dart out to grasp it. Some predators, e.g. *Notopterus, Aorichthys, Wallago* and *Xenentodon* hunt by sight, whereas the noctural fishes rely largely on smell, taste and touch and probably also on the lateral line sense organs to locate and catch the prey, as for example, *Glyptothorax, Glyptosternum, Clupisoma, Mystus, Ompok, Eutropiichthys*, etc.

In grazers the actual taking of the food is by bites, often by individual small ones. Sometimes organisms are taken singly or at other times in small groups in a rather continual type of browsing. Grazing is a characteristic of several fishes that feed on plankton or benthonic organisms. Examples are *Labeo dero, Crossocheilus*, Cobitid loaches, *Colisa fasciatus, Glossogobius, Schizothorax*, etc. A very specialised kind of grazing is found in catfishes of the family Schilbeidae, in which fish browse on parts of one another. Scales are plucked off from the body of other fishes and ingested (lepidophagy) and constitute the bulk of food of the Schilbeid catfishes. Lepidophagy is also frequent among the trouts (*Salmo* and *Salvelinus*).

In strainers there is straining of organisms from the water; this is a generalized type of feeding, as the food objects are selected by size and not by kind. Plankton filtration and swallowing of the pea-soup- like concentrate illustrates this type of feeding. A principal adaptation of the strainers is the development of numerous, closely-placed and elongated gill rakers. Examples of strainers are : *Cyprinus, Labeo dero, Garra, Puntius*, etc.

Suckers suck up food or food-containing material into the mouth. This method is practised by benthophages such as *Labeos, Garra, Cirrhinus, Crossocheilus, Noemacheilus, Betia, Glossogobius*, etc., which possess inferior mouths and sucking lips. Fishes that suck in mud to extract the organisms contained in it separate the food items from the sediment before swallowing; in others (e.g. Schilbeids, and Silurids) remains of flocculent bottom deposits can be found in the digestive tract together with high concentrations of bottom organisms.

Parasitism is one of the most unusual and highly evolved feeding habits among animals. While common examples are provided by the lampreys (*Petromyzon*) and hagfishes (*Myxine*), we do not find any representatives of this type of feeders in Jammu fishes.

REFERENCES

Das, S.M. & Moitra, S.K. 1955 : On the correlation between fish food and fish gut in fishes of Uttar Pradesh. *Proc. 42nd. Ind. Sci. Congr.*, Part IV : 20.

Das, S.M. & Moitra, S.K. 1956 - A: Studies on the food of some common fishes of Uttar Pradesh, India : Part II : On the types of fish food and the variations in the relative length of the alimentary canal with a description of the latter. *Proc. Nat. Acad. Sci.*, 26 B. Part IV : 213 -223.

Das, S.M. & Moitra, S.K. 1956.B: Studies on the food of some common fishes of Uttar Pradesh, India. Part III : Comparative anatomy of the alimentary tract and its modifications in relation to its feeding habits. *Ibid.*, 26 B. Part IV : 224-33.

Das, S.M. & Moitra, S.K.1956. C : The Surface-, the mid- and the bottom-feeding fishes of Uttar Pradesh, India. *Proc. Ind. Sci. Congr.*, Part III : 307.

Das, S.M. & Moitra, S.K. 1958 : On the feeding types of fishes and the variation in the alimentary canal in relation to food. *J. Ichth.* (*Moscow*), No. 10: 29-40.

Das, S.M. & Moitra, S.K. 1963 : Studies on the food and feeding habits of some fresh water fishes of India, Part IV : A review on the food and feeding habits with general conclusions. *Ichthyologica*, II, No. 1-2 : 107-115.

Das, S.M. & Nath, Surendra 1965 : The comparative anatomy of the alimentary tract and its modifications in relation to the food and feeding habits in some fishes of Jammu Province (India). *Ichthyologica*, 4 (1-2) : 63-78.

Das, S.M. & Nath, Surendra 1966 : The Ichthyofauna of Jammu (India) with their ecology. *Proc. 53rd. Ind. Sci. Congress*, III : 374-375.

Das S.M. & Nath, Surendra 1967 : Comparative functional anatomy of the alimentary tract in some Kashmir fishes. *Kashmir Science*, 4 (1-4) : 38-53.

Das S.M. & Nath, Surendra 1973 : On the guts of some herbivorous, omnivorous and carnivorous fresh water teleosts from Kashmir valley. *Proc. 60th Ind. Sci. Congress*, III : 548.

Lagler, Karl, F. Bardach, John E. and Miller, Robert R. 1962 : *Ichthyology— The Study of Fishes*. John Wiley & Sons, New York.

Nath, Surendra 1976 : On the incidence of stenophagy in *Glyptothorax kashmirensis* Hora (Siluriformes; Sisoridae). *Matsya*, 2 : 79.

Nath, Surendra 1978 : On the food, feeding habits and alimentary tract of *Garra gotyla gotyla* (Gray) (Cypriniformes; Cyprinidae; Garrinae). *Ibid*; 4 : 75 -76.

Nath, Surendra 1979 : On the food, feeding habits and alimentary canal of *Crossocheilus latius diplochilus* (Heckel) in Kashmir Valley and adjacent areas, *Ibid.*, 5 : 74 -76.

Nath, Surendra 1982 : On the food, feeding habits and alimentary canal of *Rasbora daniconius* (Ham.) (Cypriniformes; Cyprinidae, Rasborinae). *Ibid.*, 8 : 64 -67.

Nath, Surendra 1984 : On the food, feeding habits and alimentary canal of a hill stream teleost - *Barilius vagra* (Ham.) (Cyprini formes,Cyprinidae, Bariliinae). *J. Zool. Soc. India*, 36 (1&2) : 121-124.

Nath, Surendra 1985 : On the food, feeding habits and alimentary canal of *Glyptothorax lineatus* (day) (Siluriformes; Sisoridae) — a torrential stream teleost from Poonch Valley (Jammu and Kashmir State). *Matsya*, 11 : 89 -92.

Nath, Surendra 1986 : Observations on the general ecology of fishes of Poonch Valley (Jammu and Kashmir State) India. *Livestock Adviser*, 11 (12) : 31-36.

Nath, Surendra 1988 : On the food, feeding habits and alimentary canal of two cobitid loaches of the genus *Noemacheilus* van Hasselt from Poonch Valley (J & K State). In : *Recent Advances in Fish Ecology, Limnology and Eco-conservation*, pp. 69-72.

Nath, Surendra 1989: Studies on the guts of some Herbivorous, Omnivorous and Carnivorous freshwater teleosts from Jammu and Kashmir (India), Part - I : Anatomy. *J. Zool. Soc. India*, 40 (1&2) : 1-12.

Rec.Adv. Fish Ecol. Limn. Eco-conserv.
III: 71-78

9

Comparative Fecundity of Some Kashmir Teleosts

BANSI L. KAUL

INTRODUCTION

Although fecundity has a direct bearing on the population dynamics of a species, yet in India it has not received the deserved attention of workers. Fecundity in teleostean fishes has been studied by Raitt (1932), Kulkarni (1943), Svardson (1949), Qasim and Qayyum (1963), Das and Kaul (1965), Speare (1965), Hart (1967), Foltz (1976), Healey and Dietz (1984) and De martini (1990). The fish species selected for this study are :

1. *Schizothorax esocinus* (Heckel)
2. *Cyprinus carpio specularis* Linn.
3. *Oreinus plagiostomus* (Heckel)
4. *Crossocheilus latius diplocheilus* (Heckel)
5. *Botia birdi* Choudhary
6. *Glyptothorax kashmirensis* Hora

MATERIALS AND METHODS

The fish used in this study were collected from different lakes and streams as well as river Jhelum in Kashmir Valley. Catches were made during the spawning season using gill nets, and only gravid females were selected for the study. Total net weights were determined to the nearest 0.1 gm and total standard lengths measured to the nearest mm. Ovaries were removed in tact, blotted to remove excess water, weighed to the nearest 0.1 gm and preserved in 10% formalin for estimation of fecundity. The formalin was allowed to drip out after which cotton was used to remove any additional amount of formalin. The total fecundity was calculated by random sampling of mature ova.

Observations

Fecundity varies considerably in different fishes. Egg number in the ovaries of a fish depends upon the size and age of the fish.

1. *SCHIZOTHORAX ESOCINUS* (Heckel)

Fecundity is low in this species and varies with the age, size and weight of the fish (Table I). Fecundity ranges from 4275 in a fish measuring 14.0 cms. and weighing 100.500

gms. to 82015 in a fish measuring 56.1 cms. and weighing 1940 gms. The females measuring between 20 to 30 cms. contain ova below 12240 in number ; those measuring from 30-40 cms. have upto 25000 ova; the ones measuring from 40-41 cms. have 25000 to about 31000 ova and the females measuring above 50 cms. have more than 80000 ova in their ovaries. The over all average fecundity comes to 22737 for an average length of 31.8 cms. and body weight of 541.2 gms. The number of ova per gm. of body weight is 42.

2. *OREINUS PLAGIOSTOMUS* (Heckel)

Fecundity is very low in this species and varies with the age, size and weight of the fish (Table II). Fecundity ranges from 2600 in a fish measuring 21.6 cms. and weighing 150 gms. to 16605 in a fish measuring 33.1 cms. and weighing 515 gms. Majority of the females, measuring from 32-37 cms., contain 11139 to 16605 ova in their ovaries. The over-all average fecundity comes to 11675 for an average length of 32.9 cms. and 537.1 gms. body weight . The number of ova per gm. of body weight is about 22.

3. *CYPRINUS CARPIO SPECULARIS* Linn.

Fecundity is quite high in this fish and varies with age, size and weight (Table III). It ranges from 27000 in a fish measuring 14 cm. and weighing 223 gms. to 3,92,000 in a fish measuring 37 cms. and weighing 905 gms. The largest number of ova are possessed by such of the mature females which measure above 30 cms. in length. The females measuring below 15 cms. contain less than 30000 eggs, those between 15 cms. and 30 cms. contain between 30000 and 100000 eggs and females which measure above 30 cms. as a rule contain more than 100000 eggs. The over-all average fecundity comes to 159675 for an average length of 27.7 cms. and a weight of 518 gms. The number of ova per gm. of body weight is 308.

4. *CROSSOCHEILUS LATIUS DIPLOCHEILUS* (Heckel)

Fecundity is quite high in this fish considering its size (Table IV). It varies with size, age and weight and ranges from 3340 in a specimen measuring 6.5 cms. and weighing 6.1 gms. to 6160 in a specimen measuring 11.2 cms. and weighing 13 gms. There is a fluctuation of fecundity and sometimes a smaller specimen may have more ova than a larger specimen. In general, specimens measuring below 10 cms. have less than 5000 ova in their ovaries. The over-all average number of ova is 4502 in a specimen measuring 8.7 cms. and weighing 8.3 gms. the number of ova per gm. of body weight is 542.

5. *BOTIA BIRDI* (Choudhry)

Fecundity is very high in this species taking its size into consideration (Table V). It varies according to size, age and weight of a fish. Fecundity ranges from 19550 in a fish measuring 12.2 cms. in length and 24.1 gms. in weight to 9000 in a fish measuring 6.5 cms. and weighing 10 gms. Mature females measuring below 8 cms. have ova numbering less than 10000, those measuring from 8 cms. to 9 cms. contain 10000 to 12000 ova in their ovaries and those measuring from 9.5 to 9.8 cms. have fifteen to sixteen thousand eggs in their ovaries. Maximum number of ova are present in the ovaries of specimens measuring above 11.5 cms. numbering 16000 to 19550. The number of ova per gm. of body weight is 1025.

Table I : Morphometric Data of *Schizothorax esoclnus* (Heckel) Together with Ovarian Data and Fecundity

S. NO.	*Standard length in cms.*	*Body weight in gms.*	*Average length of ovaries in cms.*	*Weight of ovaries in gms.*	*Fecundity*
1.	53.5	1617.0	28.0	223.0	81618
2.	56.1	1940.0	27.0	235.0	82015
3.	41.5	795.500	18.0	52.500	30030
4.	33.0	377.500	13.0	32.500	16250
5.	41.0	719.700	19.0	55.300	30968
6.	22.8	125.0	10.5	10.0	6000
7.	23.2	139.600	11.0	10.400	5980
8.	16.5	109.800	7.5	10.200	4794
9.	18.4	120.700	9.1	11.300	5311
10.	14.00	100.500	6.1	9.500	4275
11.	20.5	121.0	10.0	9.0	3960
12.	30.0	351.0	12.0	24.0	12240
13.	32.0	358.800	13.5	31.200	15631
14.	35.0	462.0	14.5	38.0	19988
15.	40.0	780.0	17.5	50.0	25000
	Average = 31.8 cms.	Average = 541.2 gms.			Average = 22937

Table II : Morphometric data of *Oreinus Plagiostomus* (Heckel) together with Ovarian Data and Fecundity

S. No.	*Standard length in cms.*	*Body wt. in gms.*	*Average length of ovaries in cms.*	*Weight of ovaries in gms.*	*Fecundity*
1.	36.0	575.00	12.5	36.980	11686
2.	36.5	795.14	16.7	31.240	7716
3.	29.3	450.06	15.0	29.510	7968
4.	22.0	165.00	10.5	7.900	3950
5.	21.6	150.00	9.4	7.740	2600
6.	33.1	515.00	16.0	30.750	16605
7.	37.0	825.00	16.7	53.630	14212
8.	35.2	550.00	17.2	41.820	12713
9.	35.8	570.00	17.5	40.140	11139
10.	33.0	510.00	16.0	30.700	16578
11.	35.0	545.00	17.0	41.750	12555
12.	32.0	490.00	15.5	29.200	15651
13.	36.0	820.00	17.0	53.100	13806
14.	36.0	570.00	12.3	36.520	11686
15.	34.0	525.00	16.0	30.600	16298
	Average = 32.9 cms.	Average = 537 gms.			Average = 11675

Table III : Morphometic data of *Cyprinus carpio specularis* Linn. together with ovarian data and fecundity.

S. No.	*Standard length in cms.*	*Body Wt. in gms.*	*Average length of ovaries in cms.*	*Weight of ovaries in gms.*	*Fecundity*
1.	37.0	905.0	17.0	320.0	392000
2.	32.0	650.0	13.5	185.0	201650
3.	28.0	385.0	12.0	90.0	107100
4.	22.0	300.0	8.2	35.0	37800
5.	14.5	217.500	4.5	27.500	27750
6.	14.0	223.0	4.0	27.0	27000
7.	20.0	275.0	7.5	30.0	31500
8.	24.0	338.0	9.0	70.0	73500
9.	26.5	344.0	9.5	81.0	85680
10.	28.5	405.0	12.0	95.0	107250
11.	30.2	580.0	13.1	180.0	128800
12.	32.5	666.0	13.5	190.0	209000
13.	35.0	750.0	15.2	250.0	277500
14.	36.0	860.0	16.0	290.0	319000
15.	36.5	877.0	17.0	308.0	369600
	Average = 27.7 cms.	Average = 518 gms.			Average = 159675

Table IV: Morphometric data of *Crossocheilus latius diplocheilus* (Heckel) together with ovarian data and fecundity

S. No.	*Standard length in cms.*	*Body weight in gms.*	*Average length of ovaries in cms.*	*Weight of ovaries in gms.*	*Fecundity*
1.	11.2	13.00	4.0	4.00	6160
2.	8.0	6.500	3.2	2.500	3750
3.	9.2	9.600	3.4	3.100	4670
4.	7.7	6.600	3.1	2.550	4010
5.	7.5	7.00	3.1	2.300	3540
6.	7.8	6.500	3.1	2.500	3850
7.	9.1	9.500	3.3	3.000	4800
8.	6.5	6.100	2.9	2.100	3340
9.	10.5	11.100	3.5	3.500	4800
10.	7.9	6.700	3.2	2.500	3750
11.	10.6	11.500	3.7	3.700	5830
12.	8.7	7.100	3.8	3.200	5000
13.	8.8	7.000	3.3	3.500	5780
14.	7.2	6.100	3.0	2.210	3510
15.	9.5	9.600	3.5	3.100	4740
	Average = 8.7 cms.	Average = 8.3 gms.			Average = 4502

Table V : MorphomeTric data of *Botia birdi* Choudhry together with ovarian data and fecundity.

S. No.	*Standard length in cms.*	*Body weight in gms.*	*Average length of ovaries in* cms.	*Weight of ovaries in gms.*	*Fecundity*
1.	8.0	11.0	3.0	4.100	12210
2.	8.7	12.0	3.5	3.700	112225
3.	9.7	12.0	4.0	5.250	16000
4.	12.1	18.200	5.2	5.550	16000
5.	11.5	18.0	5.0	6.100	18010
6.	12.0	22.200	5.0	6.200	17560
7.	12.5	24.100	5.5	6.800	10550
8.	7.5	10.500	3.0	3.250	9250
9.	6.5	10.0	2.8	3.0	9000
10.	9.8	12.100	4.0	5.0	15400
11.	8.7	11.900	3.8	3.500	10460
12.	8.8	11.500	3.5	3.500	10550
13.	9.7	13.400	4.2	5.500	15200
14.	9.5	12.700	4.5	5.500	15640
15.	7.2	10.0	3.0	2.800	9610
	Average= 9.5 cms.	Average= 13.96 gms.			Average = 14311

Table VI : Morphometric data of *Glyptothorax Kashmirensis* Hora together with ovarian data and fecundity

S. No.	*Standard length in cms.*	*Body weight in gms.*	*Average length of ovaries in cms.*	*Weight of ovaries in gms.*	*Fecundity*
1.	10.5	14.440	3.1	2.440	1103
2.	10.6	14.620	3.2	3.395	1392
3.	7.7	7.610	2.4	1.135	692
4.	7.9	9.600	2.5	1.352	878
5.	6.8	8.340	2.8	2.555	1148
6.	9.4	12.200	3.0	2.220	990
7.	10.5	14.240	3.1	2.420	1113
8.	10.5	14.525	3.1	2.460	1107
9.	10.4	13.340	3.1	2.400	1080
10.	9.5	12.700	3.0	2.220	977
11.	8.7	11.605	2.8	2.210	994
12.	9.0	10.850	3.0	2.200	990
13.	10.2	13.495	3.1	3.000	1280
14.	10.5	15.880	3.0	3.100	1271
15.	8.7	10.895	3.0	2.310	1038
	Average= 9.4 cms.	Average= 12.29 gms.			Average= 1070

6. *GLYPTOTHORAX KASHMIRENSIS* (Hora)

Fecundity is not so low in this species considering its size, and varies according to size, age and weight (Table VI). Fecundity ranges from 692 to 1392 in the females measuring from 6.8 to 10.6 cms. and weighing from 7.61 to 15.88 gms. From Table VI it is evident that in the fishes with a body length above 10 cms. the fecundity ranges from 1080 to 1394 and in those below 10 cms. of length it varies from 692 to 1148. The overall average fecundity comes to 1070 eggs for an average length of 9.4 cms. and a weight of 12.29 gms. The number of ova per gm. of body weight is about 88.

Discussion

Fecundity in fishes is an important subject because of its bearing on fish production and fisheries. It depends not only on the species of fish but also on environmental factors. In nature's balance such of the fishes, in which fertilization is external, and have to face a high mortality at various stages of life history, produce large number of ova, while those which produce a smaller number of eggs are either viviparous or have some kind of parental care.

Fecundity in many species of fish is enormous. Norman and Greenwood (1963) reported that *Molva* produces as many as 28,361,000 eggs; *Gadus* 6,652,000; *Clupea harrengus* 21,000 to 47,000; and *Salmo* about 14,000 eggs per female per season on the average. In these fishes there is no parental care. Compared with these the stickle backs (*Eucalia* and *Gasterosteus*), which make elaborate nests and then provide parental care to the developing fish, the number of ova ranges from 30 to 100 per female per season. In viviparous fishes like guppies (*Lebistes*) the usual brood number is less than two dozen (Lagler et al 1963). Kyle (1926) has reported the occurrence of 14 to 60 million eggs in the ling, 8 to 10 million in the conger and 160 million in the Sunfish.

The average total fecundity of full grown *Schizothorax esocinus* for an average body length of 31.8 cms. and average body weight of 541.2 gm. is 22,937. The average total fecundity of full grown *Oreinus plagiostomus* for an average body length of 32.9 cms. and average body weight of 537 gms., is 11,675. The average total of half grown *Cyprinus carpio specularis* for an average body length of 27.7 cms. and average body weight of 518 gms. is 1,59,675. The average total fecundity of *Crossocheilus latius* for an average body length of 8.7 cms. and an average body weight of 8.3 gms. is 4,502. The average total fecundity of *Botia birdi* for an average body length of 9.3 cms. and average body weight of 13.96 gms. is 14,311. The average total fecundity of *Glyptothorax kashmirensis* for an average body length of 9.4 cms. and average body weight of 12.29 gms. is 1,070.

The comparative fecundity of all the species studied by the present author, that is the number of eggs produced per gm. of body weight, is as follows :

1. *Schizothorax esocinus* — 42
2. *Oreinus plagiostomus* — 22
3. *Cyprinus carpio specularis* — 308
4. *Crossocheilus latius diplocheilus* — 542
5. *Botia birdi* — 1025
6. *Glyptothorax kashmirensis* — 87

From the above it is clear that the highest comparative fecundity is of *Botia birdi*, followed by *Crossocheilus latius*, *Cyprinus carpio*, *Glyptothorax kashmirensis*, *Schizotho-*

rax esocinus and *Oreinus plagiostomus*, in the descending order. Thus *Botia, Crossocheilus* and *Cyprinus* are not only competing with the other endemic fishes but have now got a strangle-hold on *Schizothorax, Oreinus* and *Labeo*.

It would rather appear strange that small fishes like *Botia* and *Crossocheilus* should produce such a large number of eggs as compared to larger fishes like *Schizothorax, Oreinus* and *Cyprinus*. This is due to the much smaller size of eggs of *Botia* and *Grossocheilus*. The study of the breeding habits of these fishes has revealed that *Botia* which is essentially a weed fish living in the lakes, comes to breed in the river Jhelum in the months of July and August. The months of July and August are warm and the temperature of the lakes is higher. The temperature of the river Jhelum is comparatively lower, especially at the bottom where these fishes spawn on sand and gravel. The high comparative fecundity of *Botia* allows its off-springs to compete with other fishes in the river and during their migration to the lakes.

Crossocheilus latius produces eggs which are almost half in number per gm. of body weight as compared to that of *Botia birdi*. It is also a weed fish found in the lakes and for spawning it goes to nearby streams to spawn on sandy beds. In case sandy beds are not available it breeds among the submerged plants in the lakes. The eggs hatch at a comparatively higher temperature in this fish. The high fecundity allows it to compete successfully with the other fish species of the lakes.

Cyprinus carpio is also a prolific breeder. It breeds in the warm waters of lakes and ponds during June to September. The number of eggs produced by it is much higher than that of *Schizothorax* and *Oreinus*. This fish was introduced in the waters of Kashmir in 1956. In the last fifteen years it has spread to almost all the lakes, streams, ponds and even rice fields of Kashmir valley and outgrown in numbers all the endemic fish species. Das and Subla (1963) reported that "In any fish-catch the fisherman obtains 75% of mirror carp and about 25% of all the other fish species put together", which has now risen to 85%.

There is thus a danger of extermination of our endemic fish fauna, especially *Schizothorax* and *Oreinus*, in Kashmir valley by the mirror carp in future. The danger to *Botia* and *Crossocheilus* is much less due to their higher comparative fecundity. There is also little danger to the hill-stream fishes *Glyptothorax* and *Glyptosternum* due to the fact that *Cyprinus* has not invaded the upper reaches of mountain streams so far, perhaps because of "being more specifically adapted to flat-land living conditions" (Nikolskii 1961). The absence of the mirror carp in the mountain streams may be due to its failure to establish itself in fast currents, and if that continues to be so, we can be sure that there is no danger of annihilation to our hill-stream fish fauna as yet.

Glyptothorax kashmirensis comes down to breed in the river Jhelum during the months of June and July and spawns on sandy beds and possibly among the roots of partly submerged willows. In this respect it resembles the minnow (*Notropis*) which is reported to spawn beneath the loose bark of submerged logs (Pfleiger 1965). The fecundity is higher than that of *Schizothorax* and *Oreinus* because the offsprings of *Glyptothorax* have to migrate back to the hill-streams and face the hazards of nature. They do not spawn in the hill-streams because of the possibility of the eggs, larvae and fry being swept away by strong torrential waters. While the fingerlings and young of *Schizothorax* and *Oreinus* have to go down - stream, those of *Glyptothorax* have to go up-stream.

Fecundity of any species of fish depends not only upon the size and age of the fish, but also on the size of the egg. In such of the fish in which eggs are smaller in size, viz.,

Crossocheilus and *Botia*, fecundity is higher than those in which the eggs are larger viz., *Schizothorax, Oreinus,* and *Glyptothorax.* In *Botia* eggs are much smaller than those of *Glyptothorax*, although both the fishes are almost of the same size. Similarly the eggs of *Cyprinus* are also smaller than those of *Schizothorax* and *Oreinus*, of the same size and age. This aspect seems not to have received attention of workers. Larger eggs are richer in yolk content and produce larger larvae, which have better chances of survival than others. This view is substantiated by the fact that although *Botia* and *Crossocheilus* produce a large number of eggs, their numbers are not so large as one should normally expect. Dahl (1912) wrote in this connection "young from large eggs already from the hatching may have a larger size, which under similar environmental conditions might give a better growth" (while the present author feels that it has a much better survival value).

Svardson (1949) explained the occurrence of large number of eggs in fishes, by saying that there was a high mortality rate among the eggs and fry. He discounted the general belief that a large number of eggs remained unfertilized and consequently failed to develop. There is no doubt that a majority of eggs are lost owing to oxygen shortage during incubation, fungi and predators. He is also of the opinion that larger larvae produced from larger eggs have a better chance in natural selection than smaller larvae produced from smaller eggs.

REFERENCES

Das, S.M & Koul, B.L. (1965) : The gross anatomy of gonads and fecundity of four fishes from Kashmir. *Kashmir Sci.*, 2 (1-2) : 64-76.

De Martini E.E. (1990): Annual Variations in fecundity, egg size and conditions of the Plainfin Midshipman (*Porichthys notatus*). *Copeia* (3) : 850 - 885.

Foltz J.W. (1976): Fecundity of the Slimy Sculpin, *Cottus cognatus*, in lake Michigan. *Copeia* (4) : 802-804.

Hart J.L. (1967): Fecundity and Length-weight relationship in Lingood. *J. Fish. Res. Bd. Canada*, 24 (11).

Healey M.C. and Dietz. K. (1984): Variation in fecundity of Lake Whitefish (*Coregonus clupeiformis*) from Lesser Slave and Utikama lakes in Northern Alberta. *Copeia* 1 : 238-241.

Kulkarni C.V. (1943): Breeding habits and early stages of gourami, *Osphronemus goramy. J Bombay Nat. Hist. Soc.*, 44 (2) : 232-243.

Qasim, S.Z. and A. Qayyum (1963) : Fecundities of some fresh water fishes. *Proc. Nat. Inst. Sci. India, B*, 29 (4) : 373-382.

Raitt, D.S. (1932) : Fecundity of the Haddock. *Fish Scotland Sci. Invest.* (1) : 1-42.

Speare, E.P. (1965) : Fecundity and egg survival of the Central Johny Darter in Southern Michigan. *Copeia*, 3 : 308-314.

Svardson, G. (1949): Natural selection and egg number in fish. *Inst. Freshwater Research Drottingham, Rept.* 29 : 115-122.

Rec. Adv. Fish Ecol. Limn. Eco-conserv.
III : 79-82

10

Studies On The Bioecology of Fishes of Jammu Province (Jammu and Kashmir State) India, Part III : Spawning Ecology

SURENDRA NATH

The spawning ecology of freshwater fishes of Jammu Province has received scant attention in the past. The only earlier report available on the subject is by Malhotra *et al* (1989) which deals with the breeding ecology of some Jammu fishes. The present contribution details further observations on the spawning ecology of fishes of Jammu Province.

OBSERVATIONS

The Indo-Malayan *Notopterus chitala* spawns on benthonic objects, especially on piles of wood, fallen trees and branches in the water. The eggs are large and adhere to the substratum and are protected by the male. The male, however,not only protects the deposited eggs by actively chasing away small fishes that attempt to eat them, but also cleans the eggs from the precipitated silt and, by oscillating movements of the tail, improves the respiratory conditions for the eggs. The spawning period lasts from May to July.

The snow-trouts (*Schizothorax* and *Schizopyge*) breed when the water temperature is between 18° and 21°C. Eggs are laid in shallow pools of clear streams, amidst stones and gravel, a habit almost similar to the natural breeding of real trout (*Salmo*). Most of the marinkas (snow-trouts) spawn in rivers, less frequently in lakes, on stony grounds. The eggs stick to the stones for a short period of time for fertilization, but they are soon dislodged and undergo development, usually while wedged between the stones. *Schizothorax richardsonii* spawns from mid March to the end of May in rivers and uses gravel beds for spawning in the lower reaches of hillstreams, with clear water of moderate current. Eggs are laid at a depth of 30 - 56 cms. in Jammu.

The Cyprinids, *Labeo rohita, L.calbasu* and *Cirrhinus mrigala* become ripe in May and both male and female seek shelter under weeds and aquatic plants near banks. With the onset of monsoons, the streams become flooded and the fish move onto shallow waters such as surrounding fields and nallahs for spawning. The females lay eggs which are non-floating. Depth of water in fields where spawning occurs varies from 7.5 cm to 60 cms., and the temperature of water ranges from 76° F to 98°F. The eggs are not laid at one place and at the same time, but at intervals during which the fish keep on moving along with the current of water. As the rivers are flooded first, and the fish therein are the first to spawn, those of the tributaries wait for heavy rains which flood them sufficiently to overflow into the fields where spawning takes place. The breeding time, thus, lasts from the beginning of July to the middle of August. The eggs either sink to the bottom or rest on grass, attached by their gelatinous coats. *Labeo dero* does not migrate upstream for spawning and breeds in the cold and deeper parts of rivers.

For breeding, the Mahseers (*Tor putitora, T. khudree* and *T. mosal*) prefer small slow-flowing rivers where water is clear, well-oxygenated and the bed is sandy or covered over with gravel or pebbles. These fishes have varying spawning seasons, but the July-August period seems to be common for most of them and is regarded as the peak period. Spawning in rivers has not yet been observed, but in the water of Bhim Tal in Kumaon males and females are reported to gather together in shallow marginal areas where, amid the splashing of water, eggs are laid attached to rocks in upstream freshets and fertilized.

The Mahseers differ from other members of the Cyprinidae in as much as they are migratory in their habits and traverse long distances to get into hilly tracts for spawning purposes. They ascend to these hilly streams to find fresh feeding grounds best suited to the strength of their fry when hatched. The eggs are laid in the gravel and sand; these are sticky, adhering to the debris.

The minnows (*Chela, Salmostoma, Barilius, Danio, Chanda* and *Puntius*) breed during the early part of the monsoon on the margins of ponds, lakes and rivers. *Danio* (*Brachydanio*) *rerio* spawns readily, usually in the early morning, scattering its eggs amongst the weeds. The adults have a tendency to eat their own eggs. The hill-trouts (*Barilius*) spawn in the marginal streams in the early part of the monsoon; the fertilized eggs sink to the bottom.

Crossocheilus latius spawns amongst submerged plants and roots of trees in shallow waters of slow-moving streams, canals feeding paddy fields and in shallow pools of water. The spawning season extends from June to middle of July.

Amongst the Cobitids *Botia* spawns in rivers and streams on sandy beds. The loaches of the genus *Noemacheilus* lay their eggs either on stones and sand or on vegetation. The eggs of all loaches are benthonic.

The Sisorids (*Glyptothorax* and *Glyptosternum*) spawn on sandy beds of hill-streams and rivers. Spawning commences in early July and lasts till late August. Eggs are sticky. *Glyptosternum* spawns in summer; the eggs are large and enclosed in egg cases. *Heteropneustes fossilis* breeds during the rainy season; the yellowish eggs are laid in depressions hollowed out by the parents who guard them until hatching.

In Ophiocephalids, formation of nests, parental care and absence of floods play a great part in protecting the small number of eggs, which are laid before the rains set in. Large murrels (*Channa marulius*) breed in February and March. They build a simple nest by collecting pieces of aquatic weeds and arranging them together in a small circle. After eggs (which are floating and non-adhesive) have been laid among the weeds and fertilized, the parents fan them by flapping their pectoral fins under the nest. The smaller species (*C. punctatus*) breed from mid-April to the end of July. In *C. orientalis* pairing occurs in clear, but stagnant water, without any nest or receptacle. The eggs, after being laid, rise to the surface, spread out and float. In *C. striatus* both parents prepare a nest by biting off vegetation in a roundish area in shallow water near the margin of a canal. The eggs float at the surface in this cleared area, hatching in about three days.

The Indian glass fish (*Chanda*) spawns in thickets of fine-leaved water plants.

Badis badis lays eggs on the underside of inverted tiles or any other broken earthenware, in holes under stones, etc., as well as holes in the gravel. The eggs fall to the bottom of the lair. The male guards the eggs during incubation.

The freshwater goby (*Glossogobius giuris*) breeds in ponds and margins of rivers and lakes during the early part of the monsoon.

The giant gourami (*Colisa fasciatus*) produces a dense nest of glutinous bubbles for the eggs and early young.

On the basis of the features of their reproduction and development and, particularly, of the habitat in which they spawn, the fishes have been divided into the following ecological groups (Kryzanovsky 1948, 1949), which are applicable to Jammu fishes.

(1) Lithophils spawn on stony grounds, usually in rivers, in currents, or in oligotrophic lakes and the inshore parts of the sea, usually (but not always) in favourable respiratory conditions. This group includes sturgeons, Salmon, common barbel, *Schizothorax, Schizopyge, Noemacheilus, Tor, Badis*, etc.

(2) Phytophils spawn among plants, in stagnant or slow-flowing water on vegetation or drying plants. This group includes the carp, bream, perch, many labrids, Labeos, *Cirrhinus, Channa, Crossocheilus*, etc.

(3) Psammophils spawn on sand, sometimes attaching the eggs to plant roots. The eggs of psammophils often become encrusted with sand and usually develop in favourable respiratory conditions. This group includes the gudgeons, *Botia, Noemacheilus, Chela, Salmostoma, Barilius, Danio, Chanda, Puntius, Tor, Glyptothorax, Glyptosternum*, etc.

(4) Pelagophils spawn in the water column. The spawn and free living larvae develop in the water column, usually in favourable respiratory conditions. This group includes *Channa orientalis,Glossogobius*, and nearly all species of Clupeoids, gadoids and pleuronectids.

(5) Ostracophils spawn into the mantle cavity of molluscs and sometimes under the claws of crabs and other animals. The eggs usually develop under conditions not particularly favourable for respiration. The group includes the bitterlings, Rhodeinae, certain gudgeons (e.g.*Chilogobio, Careproctus sinensis*), etc. None of the Jammu fishes studied by the present author possesses such a spawning habit.

Of course these divisions do not include all the fishes, as there are many fish species which occupy intermediate positions. For example, in the smelt (*Osmerus*) the eggs spend part or sometimes the whole of the incubation period in the attached state and part of it free floating. The Black Sea bream (*Vimba*) spawns on both plants and stones. However, such intermediate forms are always closer to one or the other of the above ecological groups in their mode of development.

REFERENCES

Das, K.N.& Das Gupta, B.N. 1945. Breeding of the principal carps in Bengal. *Proc. Nat. Inst. Sci. India, 2* (3) : 324-327.

Das, S.M. 1959. Eggs, larvae and early life history of catfishes of Lucknow. *Proc. Ist. All-India Congr.* Zool., 2 : 30-38.

Ganpati, S.V. & Alikunhi, U.H.1948. Note on the spawning of carps in the river Thungabhadra in response to "off season" freshets. *Proc. 35th Indian Sci. Congr.*, III : 207-208.

Jones, S. 1949. Bibliography on breeding habits and development of fishes of inland water of India. *J. Zool. Soc. India,* 2 (2) : 109-125.

Khan, H. 1943. Spawning of carps and their spawning grounds in Punjab. *J. Bombay nat. Hist. Soc.*; 43 : 416.

Koul, B.L. 1988. Observations on the spawning seasons and spawning grounds of Kashmir teleosts. *Rec. Adv. Fish Ecol. Limn. Eco-conserv., I* : 54-57.

Kryzhanovsky, S.G. 1948. Ecological groupings of fishes and regular features in their formation. *Trans, Pacific Inst. Fish. U.S.S.R.,27.*

Kryzhanovsky, S.G. 1949. Ecological- morphological principles in the development of Cyprinid, Cobitid and Silurid fishes. *Trans. Inst. Morf. Zhiv., 1.*

Lagler, K.F., Bardach, J.E. and Miller, Robert R. 1962. *Ichthyology, the study of fishes.* John Wiley & Sons, New York.

Malhotra, Y.R.et al. 1989. Breeding ecology of some fishes from Jammu . *Proc. Natn. Seminar Fish. Environ.*, 37.

Mookherjee, H.K. 1945. Factors influencing the spawning of principal carps of India. *Proc. Nat. Inst. Sci. India, 8* (3) : 312-315.

Nilolsky, G.V. 1963. *The Ecology of fishes.* Academic Press, London.

Pathani, S.S. 1979. Studies on the ecology and biology of Kumaon mahseers, *Tor tor* and *Tor putitora* (Ham.).*Ph. D. Thesis, Kumaon University, Nainital.*

Pathani, S.S. 1988. Recent advances in the biology and ecology of mahseers *Tor tor* (Ham.) and *Tor putitora* (Ham.). *Rec. Adv. Fish Ecol. Limn. Eco-conserv., I* : 115-121.

Rec. Adv. Fish Ecol. Limn.. Eco-conserv.
III : 83-98

11

A Check-List of Freshwater Crustacea of Jammu and Kashmir State (India)

SURENDRA NATH

ABSTRACT

The present paper, which embodies the results of extensive investigations on the freshwater crustacean fauna of Jammu and Kashmir State (India), is the first comprehensive report on 107 species of freshwater crustaceans, representing 61 genera and 23 families of crustaceans.

INTRODUCTION

The State of Jammu & Kashmir stretches between 32° 17' N to 37°6' N latitudes and 73° 26' E to 80° 30' E longitudes. Morphologically, the State is divisible into three distinct regions, viz Jammu, Kashmir and Ladakh, each region having distinct physiographic features & faunal characteristics. It comprises a total area of about 222,000 sq. km. including those parts which are under the occupation of Pakistan and China. The State is bounded by Tibet in the East, China & Afghanistan in the West, while the Southern & South-eastern boundaries are formed by the Punjab and Himachal Pradesh States of India.

The state is drained by the mighty Indus, Jhelum, Kishenganga & Chenab rivers and their tributaries. Out of these, the Indus and the Chenab rise well to the north of the Greater Himalayas and pierce through the main ranges of the Himalayas. The Jhelum has its origin from the Pir Panjal mountain range and flows through Kashmir Valley and a part of Poonch Valley in Jammu Province. The Chenab drains the Southern side of Pir Panjal, which forms the greater part of Jammu Province.

The State contains several natural lakes, chief among which are the high altitude lakes of Ladakh [Pangong lake (4241 M) and Tso Moriri (4528 M)], Kashmir [Alapather (3660 M), Kounsarnag (4660 M), Gangabal (3600 M), Tar Sar (3000 M), Mar Sar (3500 M), Sheshnag (3800 M), etc.] and the lakes of lower altitudes, viz the Dal (1860 M), Wular (1938 M), Manasbal (1938 M), Anchar (1830 M), Nageen (1860 M) and Neelnag in Kashmir Valley and Mansar (662 M) and Saruinsar in Jammu region.

HISTORICAL

The fresh water crustacean fauna of Jammu and Kashmir state attracted the attention of a large number of investigators in the past. The earliest references to the carcinofauna of the State are contained in the works of Leydig (1860), Packard (1883), Bernard (1892),

Lawrence (1895), Alcock (1895-1900), Sars (1895-1901), Kemp (1911), Tattersall (1914) Daday (1915-1927), Gurney (1925), Bond (1934), Brehm (1936), Kiefer (1939), Brehm & Waltereck (1939), Linder (1941,1952), Longhurst (1955) and Hora et al (1955). The recent contributions on the carcinofauna of the region are those of Das (1966), Das et al (1969-1971), Malhotra and Duda (1970), Das (1970,1971), Tiwari (1972), Akhtar (1972), Malhotra & Jyoti (1972), Nath (1972-1986), Yusouf & Qadri (1975-1983), Qadri & Yusouf (1976-1980), Khan, Subla & Zutshi (1978), Khan & Zutshi (1980), Sehgal (1980), Yusouf, Balkhi & Qadri (1984-1986), Balkhi, Yousouf & Qadri (1987) and Yusouf (1988). However, the majority of these contributions deal with the carcinofauna of Kashmir Valley and a few with that of Ladakh area, the only exception to this being the reports of Nath (1972,1975,1979,1982) and Chowdhary et al (1988) which deal with some crustaceans of Jammu region. Inspite of all this, there is no comprehensive report on the freshwater carcinofauna of Jammu & Kashmir State available up to date, as all the afore-cited works deal with the crustaceans of the State in piece-meals.

Extensive investigations on the freshwater carcinofauna of the State were started by the present author in 1972 and continued upto the end of 1985. As a result of these studies 107 species of freshwater crustaceans, representing 61 genera & 23 families, were recorded from various regions of the State. The present contribution is the first comprehensive report on the freshwater crustacea of the State and provides a handy check-list of all the crustaceans recorded from the State up to date. It is hoped that this paper will serve as a basis for further investigations on the freshwater crustacean fauna of the State.

MATERIALS AND METHODS

Collections from various lakes, ponds and rivers were made by the author during the 1972-1985 period, mainly during summer. Plankton nets of mesh sizes 25 (64 μ) & 10 (157 μ) were used for collection. The samples were concentrated in the laboratory in a centrifuge and subsequently fixed with Lugol's solution. Fresh as well as fixed samples were examined under a stereoscopic binocular microscope for identification.

SYSTEMATIC LIST

		Species	*Distribution in J&K State (India)*
CLASS	:	CRUSTACEA	
Subclass	:	Branchiopoda	
Order	:	Anostraca	
Family	:	Branchinectidae	
		1. *Branchinecta ferox* Milne Edwards	Ladakh.
		2. *Branchipodopsis affinis* Sars	Nagmargh (Kashmir)
Family	:	Chirocephalidae	
		3. *Eubranchipus* sp.	Dal Lake (Kashmir)
Order	:	Notostraca	

Family :	Apodidae	
	4. *Triops cancriformis* (Bosc)	Dal Lake, Sopore, Handwara (Kashmir), Banihal and Poonch Valley.
Order :	Conchostraca	
Family :	Cyzicidae	
	5. *Eocyzicus Wulari* Das & Akhtar	Wular Lake (Kashmir)
Order :	Cladocera	
Suborder :	Haplopoda	
Family :	Leptodoridae	
	6. *Leptodora hyalina* Brehm	Manasbal Lake (Kashmir)
Suborder :	Eucladocera	
Super family :	Polyphemoidea	
Family :	Polyphemidae	
	7. *Polyphemus* sp.	Dal Lake (Kashmir)
Super family :	Sidoidea	
Family :	Sididae	
	8. *Sida Crystallina* (Muller)	Manasbal, Anchar and Neelnag, Dal and Naranbagh Lakes, (Kashmir)
	9. *Latona* sp.	Anchar and Neelnag Lakes (Kashmir)
	10. *Latonopsis fasciculata Daday*	Dal, Anchar and Neelnag Lakes (Kashmir)
	11. *Pseudosida bidentata* Herrick	Anchar and Neelnag Lakes (Kashmir)
	12. *Diaphanosoma brachyurum* (Lieven)	Manasbal, Anchar, Neelnag, Dal, Naranbagh, Trigam, Pashakuri and Bodsar Lakes.
Superfamily :	Chydoroidea	
Family :	Daphnidae	
	13. *Daphnia catawba* Coker	Dal and Gulmarg Lakes (Kashmir)
	14. *Daphnia similis* Claus	Jammu ponds
	15. *Daphnia longiremis* Sars	Dal Lake (Kashmir)
	16. *Daphnia laevis* Birge	Manasbal Lake (Kashmir)
	17. *Ceriodaphnia reticulata* (Jurine)	Dal, Manasbal and Anchar Lakes (Kashmir)

	Species	Locality
	18. *Ceriodaphnia quadrangula* (Muller)	Anchar Lakes, Mansar, Dal, Naranbagh, Neelnag Lakes and Jammu ponds.
	19. *Ceriodaphnia pulchella* Sars	Anchar, Neelnag, Dal, Manasbal Lakes (Kashmir), Surinsar Lake and Jammu ponds.
	20. *Ceriodaphnia laticauda* Muller	Anchar Lake (Kashmir) and Jammu ponds
	21. *Simocephalus vetulus* Schodler	Gulmarg, Manasbal and Anchar Lakes (Kashmir)
	22. *Simocephalus exspinosus* (Koch)	Anchar Lake (Kashmir)
	23. *Simocephalus serrulatus* (Koch)	Gulmarg pond (Kashmir)
	24. *Simocephalus elizabethae*	Anchar Lake (Kashmir)
	25. *Scapholeberis aurita* (Fischer)	Dal Lake (Kashmir)
	26. *Moinodaphnia sp.*	Dal, Anchar and Neelnag Lakes (Kashmir)
	27. *Moina affinis* Birge	Jammu ponds
Family :	Bosminidae	
	28. *Bosmina longirostris* Muller	Malpur Sar, Wular, Manasbal, Dal, Pashakuri, Neelnag, Alpathar, Trigam, Naranbagh and Anchar Lakes
	29. *Bosmina coregoni* Baird	Anchar Lake (Kashmir)
	30. *Bosmina cornuta* Muller	Dal Lake (Kashmir)
Family :	Macrothricidae	
	31. *Macrothrix* sp.	Anchar Lake (Kashmir)
	32. *Macrothrix laticornis* (Jurine)	Jammu ponds, Surinsar, Anchar, Pashakuri, and Bod-Sar Lakes
	33. *Lathonura* sp.	Manasbal Lake (Kashmir)
	34. *Ilyocryptus sordidus* (Lieven)	Anchar and Neelnag Lakes (Kashmir)
	35. *Drepanothrix* sp.	Gulmarg Lake
Family :	Chydoridae	
	36. *Chydorus sphaericus* Muller	Beehama spring, Dal, Wular, Anchar, Alapather, Malpur Sar, Hokarsar, Manasbal, Bod-Sar, Naran bagh, Pashakuri, Ladakh and Surinsar Lakes

Species	Distribution
37. *Chydorus faviformis* Birge	Kashmir and Surinsar Lake
38. *Chydorus gibbus* Lilljeborg	Anchar Lake (Kashmir)
39. *Chydorus latus* Sars	Anchar Lake (Kashmir)
40. *Chydorus barroisi* (Richard)	Anchar and Dal Lake (Kashmir)
41. *Alona affinis* (Leydig)	Dal and Anchar Lakes (Kashmir)
42. *Alona rectangula* Sars	Dal, Manasbal, Naranbagh, Trigam, Neelnag, Anchar Lakes, Malpur Sar (Kashmir), Ladakh and Surinsar Lake
43. *Alona guttata* Sars	Anchar Lake (Kashmir) and Surinsar Lake (Jammu)
44. *Alona monacantha* Sars	Anchar Lake (Kashmir)
45. *Alonella exigua* (Lilljeborg)	Malpur Sar, Beehama Spring, Dal and Anchar Lakes (Kashmir)
46. *Alonella excisa* (Fischer)	Malpur Sar, Dal and Anchar Lakes (Kashmir)
47. *Monospilus dispar* Sars	Anchar and Neelnag Lakes (Kashmir)
48. *Pleuroxus similis* Vaura	Beehama spring and Malpur Sar (Kashmir)
49. *Pleuroxus striatus* Schodler	Anchar Lake (Kashmir)
50. *Pleuroxus denticulatus* Birge	Anchar, and Neelnag Lakes (Kashmir)
51. *Acroperus harpae* Baird	Dal, Neelnag, Manasbal, Anchar Lake and Malpur Sar (Kashmir)
52. *Camptocercus rectirostris* Schodler	Dal, Wular, Alapather, Manasbal, Anchar, Kounsarnag, Bod-Sar Lake, Malpur Sar and Pashakuri Lake
53. *Alonopsis* sp.	Anchar Lake (Kashmir)
54. *Graptoleberis testudinaria* (Fisher)	Anchar, Neelnag, Naranbagh, Dal, Manasbal and Bod Sar Lakes, River Jhelum, Surinsar and Pashakuri Lakes

Subclass : Ostracoda

Suborder : Podocopa

Family : Cypridae

Subfamily : Cyprinae

55. *Cypris* sp.	Dal Lake (Kashmir)
56. *Eucypris* sp.	Gulmarg Lake (Kashmir)
57. *Cypricercus tuberculatus* (Sharpe)	Jammu ponds

Subfamily : Candocyprinae

58. *Physocypria posterotuberculata* Furtos	Dal Lake (Kashmir)

Subclass : Copepoda

Order : Calanoida

Family : Pseudocalanidae

59. *Senecella calanoides* Juday	Neelnag and Kounsarnag Lakes (Kashmir)
60. *Senecella* sp.	Dal Lake (Kashmir)

Family : Centropagidae

61. *Limnocalanus* sp.	Neelnag Lake (Kashmir)
62. *Pseudodiaptomus lobipes* Gurney	Lakes and ponds of Kashmir

Family : Temoridae

63. *Eurytemora* sp.	Dal Lake (Kashmir)

Family : Diaptomidae

Subfamily : Diaptominae

64. *Tropodiaptomus euchaetus* Kiefer	Manasbal Lake
65. *Arctodiaptomus* (*Arctodiaptomus*) *euacanthus* Kiefer	Lakes and ponds of Kashmir; also in Kounsarnag Lake
66. *Arctodiaptomus* (*Stenodiaptomus*) *stewartianus* Brehm	Ponds and Lakes of Jammu, Kashmir and Ladakh.
67. *Phyllodiaptomus blanci* de Guerne & Richard	Mansar and Naranbagh Lakes
68. *Acanthodiaptomus denticornis* Wierzejski	Dal, Manasbal, Neelnag, Naranbagh Lakes as well as ponds of Kashmir
69. *Acanthodiaptomus paravispineus*	Alpather Lake
70. *Acanthodiaptomus altissimus*	Alpather Lake
71. *Neodiaptomus schmacheri* (Poppe & Richard)	Naranbagh Lake

72. *Allodiaptomus mirabilipes* Kiefer	Lakes and ponds of Kashmir
73. *Diaptomus virginiensis* Marsh	Dal & Nageen Lakes (Kashmir)
74. *Diaptomus dilobatus* Wilson	Neelnag Lake
75. *Diaptomus purpureus* Marsh	Neelnag Lake
76. *Diaptomus kenai* Wilson	Neelnag Lake
77. *Diaptomus saltillinus* Brewer	Nageen Lake
78. *Diaptomus bogalusensis* Wilson & Moore	Nageen Lake

Subfamily : Paradiaptominae

79. *Paradiaptomus greeni* Gurney	Kashmir Valley

Order : Cyclopoida

Family : Cyclopidae

Subfamily : Eucyclopinae

80. *Eucyclops serrulatus* Fischer	Surinsar, Dal, Anchar, Hokarsar, Manasbal Lakes and Gulmarg Pond; Ladakh
81. *Eucyclops speratus* (Lillgeborg)	Anchar, Neelnag and Manasbal Lakes (Kashmir)
82. *Eucyclops agilis* (Koch)	Jammu Ponds
83. *Macrocyclops albidus* Jurine	Anchar, Manasbal and Neelnag Lakes (Kashmir); Ladakh
84. *Macrocyclops distinctus* Jurine	Kashmir
85. *Parcyclops affinis* (Sars)	Neelnag Lakes (Kashmir)
86. *Ectocyclops* sp.	Anchar and Neelnag Lakes (Kashmir)

Subfamily : Cyclopinae

87. *Cyclops bicolor* Sars	Dal and Nageen Lakes (Kashmir)
88. *Cyclops venustoides* Coker	Dal Lake (Kashmir)
89. *Cyclops scutifer* Sars	Dal, Anchar, Manasbal & Neelnag Lakes (Kashmir)
90. *Cyclops ladakanus* kiefer	Ladakh, Kaunsarnag, Trigam and Neelnag Lakes (Kashmir)

Species	Locality
91. *Cyclops exilis* Coker	Neelnag Lake (Kashmir)
92. *Cyclops varicans* Lilljeborg	Nageen Lake (Kashmir)
93. *Cyclops bicuspidatus thomasi* Forbes	Dal and Anchar Lakes (Kashmir)
94. *Cyclops vicinus* Uljanin	Surinsar Lake and Jammu ponds; Dal, Naranbagh, Trigam, Bod-Sar, Anchar, Neelnag and Manasbal Lakes (Kashmir)
95. *Cyclops insignis* Claus	Anchar Lake (Kashmir)
96. *Cyclops viridis* (Jurine)	Lakes and ponds of Kashmir
97. *Acanthocyclops vernalis* Fischer	Jammu and Kashmir
98. *Microcyclops varicans* (Sars)	Manasbal and Naranbagh Lakes
99. *Mesocyclops (Thermocyclops)* hyalinus (Rehberg)	Dal Lake and Manasbal (Kashmir)
100. *Mesocyclops leuckarti* (Claus)	Manasbal, Dal, Neelnag, Naranbagh and Trigam Lakes, Gulmargh ponds (Kashmir); Mansar Lake (Jammu)

Order : Harpacticoida

Family : Canthocamptidae

Species	Locality
101. *Canthocamptus robertcokeri* Wilson	Dal Lake (Kashmir)
102. *Canthocamptus* sp.	Anchar and Manasbal Lakes (Kashmir)

Family : Cletodidae

Species	Locality
103. *Cletocamptus albuquerquensis* Herrick	Nageen Lake (Kashmir)

Order : Lernaeida

Family : Lernaeidae

Species	Locality
104. *Lernaea kashmirensis* Malhotra & Jyoti	Dal Lake (Kashmir)

Subclass : Malacostraca

Order : Amphipoda

Family : Gammaridae

Species	Locality
105. *Gammarus pulex* (Linnaeus)	Kashmir Valley

Order : Decapoda

Suborder : Brachyura

Family : Potamonidae

106. *Potamon* (*Potamon*) *atkinsonianum* Woodmason — Poonch Valley (Jammu Province)

Suborder : Macrura

Family : Palaemonidae

107. *Macrobrachium* sp. — Jammu.

DISCUSSION

ANOSTRACA

This extremely ancient group of crustaceans is represented by only three genera, out of which *Branchinecta ferox* Milne Edwards is a high altitude species (hypsobiont) which occurs only in Ladakh region. *Branchipodopsis affinis* Sars and *Eubranchipus* sp. are the other common anostracan species which live at the lower altitudes of Kashmir Valley. *Eubranchipus* of Kashmir is of Palaearctic origin (Das, 1966) and is represented by a large number of species in Europe as well as North America.

NOTOSTRACA

This group of ancient crustaceans is represented by a single genus *Triops* Schrank. *Triops cancriformis* (Bosc) is found is Kashmir Valley as well as some parts of Jammu Province. This palaearctic branchiopod has an erratic distribution, being found in several widely-separated localities such as India, England, Lombardy, Munich, Prague, Padua, Breslau, Sweden, and Armenia. Evidence exists that *T. cancriformis* (Bosc) has survived unchanged for the last 200 million years. It is not, therefore, surprising that, in spite of its isolated existence in Kashmir from prehistoric times, *Triops* has not evolved into a new species (Nath 1979, 1985).

CONCHOSTRACA

This group of ancient crustaceans is represented by a single species - *Eocyzicus wulari* Das & Akhtar, 1971 which is found in the largest freshwater lake (Wular) in Kashmir Valley. No other representative of this crustacean order has been reported from Kashmir Valley or the rest of the State so far.

CLADOCERA

This order of "*modern waterfleas*" is represented by a large number of species in the State. *Daphnia* Muller, *Bosmina* Baird, *Alona* Baird, *Simocephalus* Schodler and *Chydorus* Leach are the most abundant and widely distributed Cladocerans in the State and occur in all sorts of habitats. *Chydorus sphaericus* Muller and *Alona rectangula* Sars are the species which occur in the high altitude lakes of the State as well as the lower altitudes of the Valley and the lakes and streams of Jammu Province. *Camptocercus rectirostris* Schodler has been reported from the high altitude lakes of the State but appears to be absent from Jammu Province.

Ceriodaphnia quadrangula (Muller), *C. pulchalla* Sars, and *C. laticauda* Muller find representation in Jammu as well as Kashmir region, while *Daphnia similis* Claus and *Moina affinis* Birge seem to be confined to Jammu region, as they have not been recorded from Kashmir or Ladakh regions so far. *Moinodaphnia* sp., which is found in several Kashmir lakes, is of palaearctic origin.

OSTRACODA

They form an independent branch of the Crustacea, which has been in existence since prehistoric times and has survived successfully to the present. After *Daphnia* (Cladocera) and *Cyclops* (Copepoda), the ostracod *Cypris* Muller is the most abundant Custacean in the State which has a very wide distribution. The other common genus is *Eucypris* Vavra which occurs in the high altitude ponds at Gulmarg in Kashmir Valley. *Physocypris posterotuberculata* Furtos occurs in Dal lake as well as North America. *Cypricercus tuberculatus* (Sharpe) has been recorded from Jammu but appears to be absent from Kashmir Valley and Ladakh region, and is also found in North America.

COPEPODA

This group of crustaceans is represented by a large number of genera and species in Kashmir waters and by a few examples in Jammu region. *Senecella calanoides* Juday, *Arctodiaptomus eucanthus* Kiefer, *A. stewartianus* Brehm, *Acanthodiaptomus paravispineus, A. altissimus, Cyclops ladakanus, Eucyclops serrulatus* Fischer and *Macrocyclops albidus* Jurine are the species which have been recorded from the high altitude lakes of Kashmir as well as Ladakh. *Cyclops vicinus* Uljanin is found in Jammu as well as Kashmir region, while *Eucyclops agilis* (Koch) appears to be confined to Jammu region.

Lernaea kashmirensis Malhotra and Jyoti is the only parasitic Copepod which has been recorded from the stone loach of Kashmir - *Noemacheilus kashmirensis* Hora (Cobitididae: Noemacheilinae) - obtained from Dal lake.

AMPHIPODA

This group is represented by a single species of Palaearctic origin, viz. *Gammarus pulex* (Lin.), which is very abundantly found in all types of aquatic habitats in Kashmir Valley and forms an important item of the food of fishes especially the Trout (*Salmo trutta fario* Lin.), the Mahseer [*Tor putitora* (Ham.)] and the Mountain Barbel [*Schizothorax richardsonii* (Gray)]. It abounds in all the streams and rivulets in the Valley and also in Wular lake, living under stones and feeding on dead leaves and other debris. *G. pulex* (Lin.) is very widely distributed in the rivers and lakes of Europe as well as most of the British Isles. It probably entered England relatively recently and, as it spread across that country, it drove out *G. duebeni* Lillj. because of its faster breeding rate (Nath 1986). There are no reports of this species from Jammu region so far.

Mansar Lake, Jammu

Fauna of Mansar Lake

Gorge en route to Sheshnag, Kashmir

River Jhelum, Kashmir

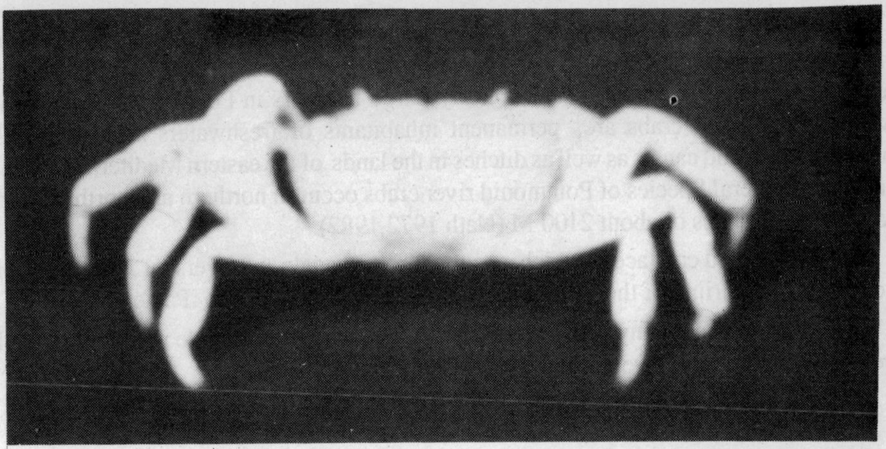

River Crab - *Potamon* (*Potamon*) *atkinsoniamum* from Poonch Valley

Boulder stream - Poonch Valley

DECAPODA

The only decapod (brachyura) freshwater crustacean reported from the State so far is the freshwater river crab—*Potamon* (*Potamon*) *atkinsonianum* Woodmason (Cyclometopa : Potamonidae)— which occurs in fairly large numbers in Poonch Valley (Jammu Province). Potamonid crabs are permanent inhabitants of freshwaters such as rivers, mountain streams and canals as well as ditches in the lands of the eastern Mediterranean and Asia Minor. Several Species of Potamonid river crabs occur in northern and north eastern India, even at altitudes of about 2100 M (Nath 1972,1982).

Another decapod crustacean which occurs in some canals and rivers of Jammu region is the freshwater shrimp of the genus *Macrobrachium* Bate (Macrura :Palaemonidae).

There is no record of any decapod crustacean from the high altitude region of Ladakh as well as the relatively lower altitudes of Kashmir region.

ACKNOWLEDGEMENTS

I am thankful to the Director, British Museum of Natural History, London as well as Dr. G.A. Boxshall and Dr. Roger J. Lincoln of the Department of Zoology, Section of Crustacea, British Museum, London for confirming the identification of some of the specimens under report. Thanks are also due to Dr. K.K.Tiwari and Dr. B.K. Tikader of the Zoological Survey of India for their valuable opinion and help in the identification of a few specimens.

BIBLIOGRAPHY

Akhtar, S. 1972. Qualitative and quantitative studies on freshwater plankton Rotifera, Cladocera, Ostracoda and Copepoda of Kashmir, *Ph. D. Thesis, University of Kashmir*.

Alcock, A. 1895-1900. Materials for a Carcinological fauna of India. *J. As. Soc. Bengal, Calcutta*.

Alcock, A. 1909. Diagnoses of new species and varieties of freshwater crabs. *Rec. Ind. Mus., 3* (4) : 375-381.

Balkhi, M.H., Yousuf, A.R. and Qadri, M.Y. 1987. Hydrobiology of Anchar Lake, Kashmir. *Comp. Physiol. Ecol., 12* (3) : 131 -139.

Bate, R.H., Robinson, E. and Sheppard, L.H. 1982. *Fossil and Recent Ostracods*. Ellis Harwood.

Bernard, H.M. 1892. *The Apodidae, A Morphological Study*. London.

Bond, R.M. 1934. Report on Phyllopod Crustacea (Anostraca, Notostraca and Conchostraca) including a revision of Anostraca from the Indian Empire. *Mem. Connecticut Acad., 10* : 29-62.

Brehm, V. 1936. Yale North India Expedition. Article XVI. Report on cladocera. *Ibid., 10*: 283-297.

Brehm, V. 1950a. Contributions to the freshwater fauna of India, Part I. *Rec. Ind. Mus., 48* (1) : 1-8.

Brehm, V. 1950 b. Contributions to the freshwater fauna of India , Part II. *Ibid. 48* (1) : 9-28.

Brehm, V. and Waltereck, R. 1939. Die Daphniden der Yale Indian Expedition *Inst. Rev. Hydrobiol. Leipzig, 38* : 1 - 19.

Chowdhary, S.K., Sharma, J.P. and Srivastava, J.B. 1988. Limnological studies of Jammu ponds. In : *Recent Advances in Fish Ecology, Limnology and Eco-conservation* : 73 - 78, Creative Publishers, New Delhi.

Daday, de Dees, E. 1915. Monographi systematique des phyllopodes conchostraces. *Ann. Sci. Nat. Zool. Ser. 9* (20) : 37 - 330.

Daday, de Dees, E. 1923. Monographi systematique des phyllopodes conchostraces. *Ibid., 10* (6) : 225 -390.

Daday, de Dees, E. 1925-1927. Monographi systematique des phyllopodes conchostraces. *Ibid., 10* (8) : 143 - 184 and *10* (10) : 1 - 112.

Das, S.M. 1966. Palaearctic elements in the fauna of Kashmir. *Nature, 212* : 1-6.

Das, S.M. 1966. Studies on freshwater plankton from Kashmir. *Proc. Indian Sci. Congress*, III.

Das, S.M. 1970. On a new species of *Apus* (*Triops*) Schaeffer 1756 (Crustacea, Branchiopoda, Notostraca) from Kashmir. *Ichthyologica, 10* (1-2) : 5-7.

Das, S.M. 1970. Ecology of Dal Lake, Kashmir, India. *Kashmir Sci.*, 7 (1-2) : 16-24.

Das, S.M. 1971. On a new species of *Apus* (Crustacea, Notostraca) from Kashmir. *Proc. 58th Ind. Sci. Congr.*, Part III : 555-556.

Das, S.M. 1989. *Handbook of Limnology and Water Pollution in India*. South Asian Publishers, New Delhi.

Das, S.M. and Akhtar, S. 1970. A report on freshwater Cladocera from Dal Lake, Kashmir. *Kashmir Sci.*, 7 (1-2) : 133-137.

Das, S.M. and Akhtar, S. 1971. On a new species of *Eocyzicus : Eocyzicus Wularin*. sp. (Conchostraca ; Cyzicidae; Crustacea). *Ibid.*, 8 (1-2) : 111-114.

Das, S.M., Choudhry, S., Ahmad, N., Akhtar, S., Peer, S. and Rashid, M. 1970. Studies on organic production in high altitude lakes of India, Part II : The Zooplankton, Phytoplankton and Pedon of the high altitude Kashmir lakes, Kounsarnag and Alpather, with correlation of Plankton Volume, pH and temperature of Dal Lake. *Ibid.*, 7 (1-2) : 119-132.

Das, S.M., Daftari, S., Singh, H. Akhtar, S., Chowdari, S. and Ahmed, N. 1969. Studies on organic production in High altitude Lakes of India, Part I : The general ecology and Zooplankton of Kashmir Lakes. *Ibid.*, 6 (1-2) : 1-22.

Fox, M. 1949. The rediscovery of *Apus* in England. *Proc. Zool. Soc. London, 119* : 693 - 702.

Gurney, R. 1925. Some Asiatic specimens of *Apus*. (i) *Apus cancriformis Schaeffer* (ii) *A. granarius* Lucas. *Rec. Ind. Mus.*, 27 : 439 - 442.

Holthuis, L.B. and Hemming, F. 1956. Proposed use of the plenary powers to validate the generic name *Lepidurus* Leach, 1819 and to designate a type species for and to determine the gender of *Triops* Schrank 1803. *Bull. Zool. Nomenclature, 12* : 67-85.

Hora, S.L., Malik, G.M. and Khajuria, H. 1955. Some interesting features of the aquatic fauna of the Kashmir Valley. *J. Bombay Nat. Hist. Soc., 53* (1) : 140-143.

Kemp, S. 1911. Notes on the occurrence of *Apus* in Eastern Asia. II. Notes on Major Walton's specimens and on others from Kashmir with a list of previous records from Eastern Asia. *Rec. Ind.Mus.*, *6* : 353 - 357.

Khan, M.A., Subla, B.A. and Zutshi, D.P. 1978. A new Crustacean for India. *Geobios, 5* (2): 81.

Khan, M.A., and Zutshi, D.P. 1980. Contribution to high altitude limnology of Himalayan System I. Limnology and primary productivity of the Plankton community of Nilnag lake, Kashmir *Hydrobiologia, 75* : 103 - 112.

Kiefer, F. 1939. Scientific results of the Yale North India Expedition : Biological report No. 19. Copepoda. *Mem. Indian Mus.*, 1939, *13* (2) : 83-203.

Lawrence, Walter R. 1895. *The Valley of Kashmir*. London.

Leydig, Fr. 1860. *Monographie der Daphniden*. Tubingen.

Linder, F. 1941. Contributions to the morphology and taxonomy of the Branchiopoda Anostraca. *Zool. Bidr. Uppsala, 20* : 101 - 302.

Linder, F.M. 1952. Contributions to the morphology and taxonomy of the Branchiopoda, Notostraca, with special reference to the N. American species. *Proc. U. S. Nat. Mus., 102* :1 - 69.

Longhurst, A.R. 1955. A review of the Notostraca. *Bull. Brit. Mus.* (N.H.) *Zoology, 3* (1) : 1-57.

Malhotra, Y.R. and Duda, P.L. 1970. A new fairy shrimp, *Branchinecta acanthopenes* n. sp. (Anostraca : Branchinectidae) from India. *Crustaceana, 18* (2) : 173-176.

Malhotra, Y.R. and Jyoti, M.K. 1972. A New copepod parasite *Lernaea Kashmirensis* n. sp. (lernaeidae, Lernaeinae) infecting stone - loach of Kashmir. *Proc. 59th Ind. Sci. Congress*, Part III : 454.

Nath, Surendra. 1972. Studies on the freshwater Crustacea of Jammu and Kashmir state (India) Part I : *Potamon atkinsonianum* Wood-Mason (Decapoda; Brachyura; Potamonidae) a new record from Poonch Valley (J&K State). *Proc. 59th Indian Sci. Congress*, III : 452-453.

Nath, Surendra. 1975. Studies on the freshwater Crustacea of Jammu and Kashmir State (India) Part II : *On Triops cancriformis* (Bosc) (Branchiopoda ; Notostraca; Apodidae) from Kashmir Valley and adjacent areas, with remarks on the taxonomic status of *Apus kashmiriensis* Das 1970. *Proc. IIIrd All - India Congr. Zool.*, 3 : 42-43.

Nath Surendra. 1979. Extension of range of *Triops cancriformis* (Bosc) (Branchiopoda ; Notostraca ;(Apodidae) to Poonch Valley (Jammu & Kashmir State) *J. Bombay Nat. Hist. Soc.* , *76* (3) : 543-544.

Nath, Surendra. 1982. Extension of range of the river-crab—*Potamon* (*Potamon*) *atkinsonianum* Woodmason - (Brachyura; Cyclometopa; Potamonidae) - to Poonch Valley (Jammu and Kashmir State). *Ibid., 79* (3) : 705.

Nath, Surendra. 1985. On the taxanomic status of *Apus Kashmiriensis* Das (Crustacea : Branchiopoda : Notostraca : Apodidae). *Ibid., 82* (2) : 424-427.

Nath, Surendra. 1986. On the occurrence of the Amphipod *Gammarus pulex* (Lin.) in Kashmir Valley, with remarks on the ecology of the species *J. Zool. Soc. India, 38* (1-2) : 115-116.

Packard, A. 1883. *A Monograph of North American Phyllopod Crustacea*. Washington

Qadri, M.Y. and Yousuf, A.R. 1977. Limnology and fisheries of Lake Manasbal I. Distribution of Zooplankton. *J.Sci.* (*Univ. Kashmir*), *4* (1-2).

Qadri, M.Y. and Yousuf, A.R. 1977. Some new records of Crustacea from Kashmir *Curr. Sci., 46* (24) : 859-860.

Qadri, M.Y. and Yousuf, A.R. 1980. Influence of some physico-chemical factors on the seasonality of Cladocera in Lake Manasbal. *Geobios,7* : 273-276.

Sars, G.O. 1895, 1899, 1900. *An account of the Crustacea of Norway*, Vol. I. Amphipoda 1895; Vol.2. Isopoda 1899 ; Vol. 3 Cumacea, 1900.

Sars, G.O. 1901. Amphipoda and Phyllopoda. On the crustacean fauna of Central Asia Pt. I. *Ann. Mus. St. Petersbourg, 6* (1): 130-164.

Sehgal, K.L. 1983. *Planktonic Copepods of Freshwater Ecosystems*. Interprint, New Delhi.

Tattersall, Walter M. 1914. Notes on some Amphipods collected on the Pamirs at an altitude of 15600 ft. *Rec. Ind. Mus., 10* (3) : 213-214.

Tiwari, K.K. 1951. Indian species of the genus *Apus*, with descriptions of two new species. *Rec. Ind. Mus., 49* ; 197 - 206.

Tiwari, K.K. 1972. Taxonomic status of two recently described Branchiopoda from Kashmir, India. *Crustaceana, 23* (3) : 311-314.

Yousuf, A.R. 1988. Copepod plankton of Lake Manasbal, Kashmir. *J. Indian Inst. Sci., 68*: 307-313.

Yousuf, A.R., Balkhi, M.H., and Qadri, M.Y. 1984. Summer and Winter cladoceran communities of the Anchar Lake, Kashmir. *Geobios new Reports, 3* (1) : 66-67.

Yousuf, A.R., Balkhi, M.H. and Qadri, M.Y. 1986. Limnological features of a forest Lake of Kashmir. *J. Zool. Soc. India, 38* (1-2) : 29-42.

Yousuf, A.R. and Qadri, M.Y. 1975. Cladocara of Malpur Sar, Kashmir. *J. Sci. Univ. Kashmir, 3* (1-2) : 87-92.

Yousuf, A.R. and Qadri, M.Y. 1981 a . Seasonal distribution of family Chydoridae (Cladocera, Crustacea) in Lake Manasbal. *J. Inld. Fish. Soc. India, 13* : 36-45.

Yousuf, A.R and Qadri, M.Y. 1981. b. Ecology of *Diaphanosoma brachyurum* Lieven (Cladocera : Crustacea) in Lake Manasbal, Kashmir. *J. Indian Inst. Sci., 63* (C) : 35-42.

Yousuf, A.R. and Qadri, M.Y. 1983. Seasonal distribution of family Sididae, Daphnidae, Bosminidae, Macrothricidae and Leptodoridae (Cladocera : Crustacea) in Lake Manasbal, Kashmir. *J. Inld. Fish. Soc. India. 15* (1-2) : 13-20.

Rec. Adv. Fish Ecol. Limn. Eco-conserv.
III : 99-102

12

A Report On The Study of Fish Ponds In Ladakh

A. WANGANEO, D.P.ZUTSHI,* AND *R. WANGANEO***

INTRODUCTION

Ladakh, the most elevated region (alt. 4000 m. a. s.l.), is situated between the geographical co-ordinates of 33-35°N latitude and 75.5 - 79°E longitude towards the north of India. The presence of mountainous terrain, rich in lime stone and the extremes of temperatures (Max. summer temp., 37°C and Min. winter temp., -30°C) are the main governing factors where most part of Ladakh is barren (70%). In order to compensate the less agricultural produce it is proper to utilise the natural resources viz. water bodies (both saline and fresh water) for developing fish so that proteinacious diet is made available to the local population throughout the year.

It was with this idea that the Ladakh region was surveyed by the present authors during 1980 summer. In the present paper, investigations carried out on the fish ponds of Ladakh region have been detailed.

Fish ponds were constructed in Ladakh and stocked with 2" size fingerlings of carp (*Cyprinus carpio spp.*) in October, 1967. The fingerlings were brought from Bandipore. Initially 2000 fingerlings were stocked and after about one year additional fingerlings (350) were added. Between 1974-78 additional fingerlings of *Schizothorax* (7500) were introduced into the ponds.

MATERIALS AND METHODS

Surface water samples were collected directly while a plexiglass Ruttner sampler was used for collecting samples from deeper region. Samples for chemical analysis were collected in polyethylene bottles of one litre capacity, while water samples for dissolved oxygen were collected in separate glass bottles 125 ml±2.5). Water analysis was carried out following the standard methods as given in Golterman and Clymo and APHA (1971).

The quantitative analysis of plankton population was done by filtering a known volume of water (1 litre) through a plankton net No. 25 (64 m µ grade). The samples were preserved with Lugol's solution. The phytoplankton results have been expressed as units 1^{-1}, while zooplankton results have been expressed as individuals 1^{-1} after Wanganeo (1980).

* Department of Limnology, Bhopal University, Bhopal , (M.P.) - 462026.

** S.V. College of Bairagarh, Bhopal, M.P., India.

RESULTS AND DISCUSSION

Five ponds, which have been constructed in Shy, were surveyed during the course of the present investigations. Three ponds were comparatively large in size (15' × 50') and had good population of *Cyprinus carpio* spp. (Pond 'A').The fourth pond had predominance of *Schizothorax* sp. (Pond 'B'). One of the ponds was used as a nursery. These ponds were infested with macrophytes, the major ones being *Potamogeton perfoliatus* and *P. pectinatus*. One of the ponds was completely neglected and was full of *Potamogeton*. Nursery ponds were also full with *Chara*.

Table 1 sums up the physico-chemical characteristics of these two ponds. The ionic composition of the water in these ponds were calcium dominated and was similar to the lakes found in Kashmir Valley at an average altitude of 1560 m.a.s.l.

Table 1 : Physico-chemical characteristics of Pond 'A' and Pond 'B' in Ladakh

Parameters		*Pond A*		*Pond B*	
		Surface	*Bottom*	*Surface*	*Bottom*
Depth	m	1.5		2.5	
Secchi depth	m	1.5		2.5	
Water Temp.	°C	15	15	15	13
Cond. µS 25°C		289	286	249	259
pH		8.1	8.1	8.4	8.2
D.O.	mgl^{-1}	8.96	4.8	11.84	11.20
Ca	mgl^{-1}	38.4	80	70.80	47.20
Mg	mgl^{-1}	30.8	11.52	6	12
Tot.Alk.	mgl^{-1}	165	155	126	131
Cl	mgl^{-1}	15	15	9	9
Sio_2	mgl^{-1}	7.8	7.4	6	6.2
Fe	μgl^{-1}	80	30	ND	ND
NO_3-N	μgl^{-1}	80.70	40.93	120.18	70.54
NH_4-N	μgl^{-1}	13.65	12.95	6.95	12.10
PO_4-P	μgl^{-1}	< 5	< 5	< 5	< 5
TP	μgl^{-1}	27	30	35	36

ND = Not detected

PLANKTON POPULATION

1. PHYTOPLANKTON

Total phytoplankton population recorded higher values in the surface waters compared to the bottom waters in both the ponds. A value of 28,228 units/1 and 10,840 units/1 was recorded for total phytoplankton population in surface and bottom water of Pond 'A' while, in Pond 'B' surface recorded 17.668 µ/1. Distribution of total phytoplankton population into various classes is given in Table 2.

Table 2. : Class wise distribution of Phytoplankton of Pond A and Pond B in Leh Ladakh

Phytoplankton Classes	*Pond A* *Surface* (Units/1)	*Pond A* *Bottom* (Units/1)	*Pond B* *Surface* (Units/1)	*Pond B* *Bottom* (Units/1)
Bacillariophyceae	27,168	10,240	7,530	8,368
Chlorophyceae	780	340	12,000	3.180
Cyanophyceae	260	260	140	1,120
Chrysophyceae	20	ND	ND	ND

The sequence of dominance of various classes is as follows :

Bacillariophyceae > Chlorophyceae > Cyanophyceae > Chrysophyceae

Bacillariophyceae was mostly represented by *Fragilaria crotonensis* (Max. population: 18,480 μ/1 in pond A and 5,270 μ/1 in pond B); and *Cymbella* sp. (Max. population 2,040 μ/1 in Pond A and 2,140 μ/1 in Pond B); and *Navicula* sp. (Max. Population 1,320 μ/1 in Pond A, and 860 μ/1 in Pond B). Chlorophyceae was mainly dominated by *Scenedesmus bijugatus* (Max. population 280 μ/1 in Pond A, and 940 μ/1 in Pond B), *Scenedesmus armatus* (Max. population 140 μ/1 in Pond A and 800 μ/1 in Pond B) and *Cosmarium* sp. (Max. population 80 μ/1 in Pond A and 1,260 μ/1 in Pond B). *Oscillatoria* contributed the main bulk in Cyanophyceae (Max. population 180 μ/1 in Pond A and 1,120 μ/1 in Pond B). Chrysophyceae was represented by *Dinobryon* only (Max. population 20 μ/1 in Pond A only).

ZOOPLANKTON POPULATION

Surface water samples were deprived of zooplankton population, while the bottom waters were rich in rotifer population in comparison to other classes of zooplankton. A max. density of 210 individuals/1 and 350 ind./1 was encountered in bottom waters of Pond A and Pond B respectively, in which rotifer population contributed 180 ind./1 in Pond A and 280 ind./1 in Pond B. The bulk of rotifer population was contributed by *Polyarthra* sp. (100 ind./1 in Pond A and 40 ind./1 in Pond B); *Keratella quadrata* (80 ind./1 in Pond A); *Monostyla bulla* (60 ind./in Pond B) and *Mytilina ventralis* and *Euchlanis dilatata*, both contributing 40 ind./1 in Pond B. Other than this zooplankton population was represented by *Cyclops* sp. from Copepods (40 ind./1 in Pond B) *Chydoras sphaericus* from Cladocerans (10 ind./1 in Pond A and Pond B) and 10-20 ind/1 from Ostracoda in both the ponds.

Zooplankton sp. encountered in the present investigation in Pond B alone were as follows :

Lecane luna, L. closterocerca, Lecane sp., *Mytilina mucronata, Lapadela ovalis, Trichocerca* sp., *Colurella* sp., *Trichotria tetractis* and *Squatinella mutica.*

FISH POPULATION

The survival rate of the introduced fish was almost 99%. The sale of the fish had increased many fold. As per the official records, $3\frac{1}{2}$ kgs were sold in 1979-80 while during 1980-81 it was $87\frac{1}{2}$ kgs. This increase in fish sale is of great significance. Besides this the availability of abundant natural food in the form of plankton population supports the survival of introduced fish variety. Further the presence of a large number of fresh as well as saline water bodies indicate that fishery can be developed on a large scale using almost all the existing water bodies.

REFERENCES

American Public Health Association, 1971, Standard methods for the examination of water and waste water. *APHA NY*.

Goltermen, H.L. and Clymo, R.S. 1969. Methods for chemical analysis of freshwater. *IBP Hand book No. 8*. Blackwell scientific Publication, Oxford.

Hutchinsen, G.E. 1957. A treatise on Limnology. Vol.I. Geography, Physics and Chemistry. *John Wiley and Sons, Inc. N.Y.* and *London*. 1015 P.

Wanganeo, A. 1980. Phytoplankton photosynthesis, Nutrient dynamics and trophic status of Manasbal lake Kashmir. *Ph. D. Thesis of Kashmir University* (un published).

Rec. Adv. Fish. Eco. Limn. Eco-conserv.
III : 103-119

13

The Ichthyogeography of Jammu Province (Jammu and Kashmir State) India

SURENDRA NATH

INTRODUCTION

The Jammu Province of Jammu and Kashmir state embraces the hilly country extending down from the snowy range of Pir Panjal mountains bounding Kashmir on the South to the plains of the Punjab. It lies between lat. 32° 20' and 33° 10' and long. 74° 45' and 75° 55' and is bounded on the North by the Chenab River and Bhaderwah, on the South by the Punjab, on the East by the hilly district of Chamba, and on the West by the Punjab and Naoshera. The Jammu Province is mainly drained by the River Uj, Tawi, Chenab, Manawar - Tawi, Poonch and the Jhelum Rivers.

The fishes of Jammu Province are quite distinct from those of Kashmir Province, the two regions being separated by the Pir Panjal range, which is also the watershed for the Jhelum as well as the Chenab drainage systems. River Jhelum flows through Kashmir Valley and a part of Poonch district of Jammu Province. River Chenab drains the southern side of Pir Panjal which forms the greater part of the Province.

In spite of the highly interesting ecological and geographical features of the region, its ichthyofauna had not at all been studied until recently. In the vast literature on the ichthyofauna of India there is no comprehensive account available on the systematics, ichthyogeography and bioecology of the freshwater teleosts of Jammu Province. The present author, therefore, started intensive studies on the ichthyofauna of Jammu Province in 1964, which were continued up to 1990, and also studied the bioecology of the fishes of the region for the first time (Das & Nath 1965 - 1971; Nath 1969-1987).

PHYSIOGRAPHY

Geomorphic Features

The Province of Jammu comprises a sub-montane tract forming the upper basin of the Chenab and situated at an average altitude of 305 to 915 masl. The Province and its Siwaliks are looked over by the Middle Himalayan mountains of Udhampur, Kishtwar and Bhaderwah, till the Pir Panjal is crossed now at a height of 2195 masl by the Jawahar Tunnel which opens into the Valley of Kashmir. Excepting its southern-most part, the Jammu Province is all hilly terrain, traversed by several rivers and streams. Besides the town of Jammu, which is situated at an altitude of 366 masl on the southern slope of Siwaliks along the right bank of the Tawi River (a tributary of the Chenab River), the Province also includes the districts of Kathua, Udhampur, Doda, Rajouri and Poonch.

In the south the Province is rimmed by an alluvial plain of the Chenab, Tawi, and Ravi rivers and their tributaries. In width it varies from seven to thirty Km. and sprawls in the southern parts of Akhnoor, Ranbirsinghpora, Samba and Kathua. This Plain constitutes a part of the Great Plains of India and resembles the landscape of Punjab. The altitude of this highly productive plain varies between 325 to 350 masl.

Being in the south of the Siwaliks, the Outer Plain of Jammu Province is traversed by the numerous hill-torrents locally called *Khads*. These torrents remain dry over the greater parts of the winter and summer seasons and attain an enormous size during the rains. With the onset of monsoons, the ravines attain the shape of gushing torrents which are unfordable. Occasionally, these flooded torrents transport massive quantities of sand, red silt, pebbles, boulders and shingles. The channels of these torrents are characterised by clumps of red growths, locally called *Khar*.

Jammu Province is made up of six districts : Jammu, Kathua, Rajouri, Poonch, Udhampur and Doda.

The Jammu district is the gateway of Jammu & Kashmir State, lying as it does on the nearest railhead of the State as a whole, as well as on the national highway from Pathankot to Uri. The district comprises four tehsils - Jammu, Samba, Ranbirsinghpora and Akhnoor.

Kathua district comprises four tehsils - Kathua, Hiranagar, Basohli and Bilawar. Major part of the district is *kandi*, the soil bearing nitrogen, phosphates and potash.

Poonch district consists of three tehsils - Haveli, Mendhar and Nowshera. A sizable chunk of the district has fallen on the Pakistan side of the Line-of Control. The area comprises about 6100 sq. Km. of hilly country, of which about 3000 sq. Km. lie within the Middle (or Lesser) Himalayan zone.

Rajouri district has only two tehsils - Rajouri and Buddhal. Rajouri is situated on the side of a low range of jungle-covered hills, about 46 m. above the right bank of the Rajouriwali Tawi river. Opposite the town on the left bank of the river, the valley opens out into a wide and extensive plain.

Udhampur district comprises the tehsils of Udhampur, Ramnagar, Reasi and Gulabargh. It falls in the region of the Outer hills, lying somewhat to the south of the Pir Panjal range which separates Jammu region from Kashmir region. The average altitude is between 610 to 1220 masl.

Doda district forms the second largest district (area-wise) of the Jammu & Kashmir State, the first being Ladakh. It comprises the tehsils of Ramban, Doda, Kishtwar and Bhaderwah. The district covers what is known as the Outer Hills region. Its altitude ranges between 600 to 1200 masl.

The Mountain Ranges

A. *The Siwaliks* : To the north of the Jammu plains for a distance of about 200 Kms. between Ravi and Jhelum rivers stretch the youngest mountains of the subcontinent known as the Siwaliks. The width of the Siwaliks is between 20 to 50 kms. and height between 600 to 1200 masl.

The slopes of the Siwaliks facing the outer plains are gentle, covered with deciduous forests, evergreen bushes and numerous scrubs. The northern slopes are steep, with thin cover of vegetation. In the Siwaliks to the East of Jammu city are situated the Mansar and Surinsar lakes at an altitude of about 600 masl.

The undulating slopes, adjacent to the Outer plains up to an elevation of 300 meters between Ravi and Chenab rivers are locally known as *kandi*. The *kandi* land is characterised by numerous torrents soils and scarcity of water. Consequently, the rate of soil erosion during the rainy season is heavy.

In between the Siwaliks and the Lesser Himalayas lie the longitudinal valleys locally known as the *Duns*. The important *duns* are Basohli, Ramkote, Dansal, Udhampur and Sunderbani.

B. *The Middle Himalayas* : They are locally known as *Pahar* in Jammu province and sprawl between the Ravi in the East and the Poonch in the West, though they continue towards northeast beyond Muzaffarabad. In elevation they vary between 1820 to 2240 masl. The Middle Himalayas have a width of about 60 Kms. in the eastern parts of Jammu province which decreases to only about 10 Kms. near Rajouri.

The Middle Himalayas in general have an east-west extension. From the ridges, however, a number of lateral ridges and spurs radiate out in all directions. At Tanger, a branch ridge runs due south-west. It is the Trikuta Mountain which has limestone strata.

Several important rivers, e.g. Tawi, Manawar-Tawi, Basanter and Uj originate in the Middle Himalayas. The rock strata of these mountains are dominated by sandstone, shale and light yellow limestone of Miocene/ Murree system and Eocene deposits.

The southern slopes of Middle Himalayas are covered by forests. In the high altitudinal zones Deodar (Oak), blue-pine and silver-fir dominate, while in the areas of low altitudes (below 1800 m.) the monsoon deciduous forest and scrubs predominate.

The Jammu - Srinagar National Highway has been carved out through the Middle Himalayas which crosses this system at Patni Top Pass at an altitude of about 1950 meters.

Drainage

The main river systems which drain the watersheds of Jammu Province are the Chenab, the Jhelum and the Ravi. The approximate stream-length of these rivers along with their tributaries is 1955 Kms. The majority of the streams in Jammu Province are seasonal and carry very little water throughout the year.

A.*The Chenab* : It rises in the Himalayan contour of Lahaul (Himachal Pradesh) as two streams, the Chandra and the Bagha. The former issues from a large snowbed on the southeast of the Baralacha 6096 masl and the Bagha on the northwest slopes of the pass. These flow more-or-less parallel to each other but the Chandra deviates to the north-west a distance of about 176 Kms., meeting the Bagha at Tandi. The united stream Chandrabagha or Chenab flows through the Pangi Valley in the Chamba State and then flows through the Padar district of Kashmir at an elevation of about 1830 masl. From Tandi to Kishtwar the fall of the river is 10.5 meters/1.6 Km. The Wadwan river system enters the Chenab at Kishtwar. Its perennial source is in the glacier of Nunkun. From Arnas the Chenab flows south and leaves the mountains at Akhnoor, 288 Km. below Kishtwar. Above Akhnoor the river-bed is about 61 ms. high and full of cliffs. From the Baralacha pass the length of the Chenab is about 640 Kms. and the total fall is 5030 ms., or 12 meters/1.6 Kms. The Chenab is a typical mountain river with rapids in its upper course, till it enters the plains at Akhnoor where the river water is used for irrigating the Bijwara zone.

After passing through the deep gorges, the Chenab comes out into the open at Ramban. From Ramban it takes a westerly course passing through Reasi tehsil of Udhampur district and Akhnoor tehsil of Jammu district. Before entering Pakistan, it gets divided into several channels near Akhnoor. The river-bed is rocky at many places and big boulders get displaced and are carried down by the rapid current.

The principal tributaries of the Chenab are the Tawi, the Anji, the Pahi, the Neeladhab and the Rajouriwali Tawi. The Tawi has a stream-length of about 275 Kms; its main side-streams are Naddar, Duddar and Jhajjarkotli.

The Tawi, which is the chief fishery resource at Jammu, originates from the spring Kailashkund, situated atop the mountain Sewaj in Ramnagar tehsil (in about lat. 32° 53' and long. 75° 34'). It course is first northwest toward Chinaini for about 42 Kms; in this part the river forms along among large boulders, confined by rocks which are capped with a deposit of alluvium of rounded blocks of gneiss embedded in the earth. These banks are in part clothed with shrubs and trees that have found a footing upon them.

From Chinaini it bends to the southwest for a distance of about 24 Kms., the valley being still bounded by high hills. It then emerges upon a more open country, through which it flows in a southerly direction for some 16 Kms. , when it meets a range of hills through which it pierces, and then takes a westerly course, winding very considerably; it then pierces another range of hills and reaches the town of Jammu, 29 Kms. from which it joins the Chenab near Chaprad in the tehsil Ranbirsinghpora. Its total length is 112 Kms. At Jammu, the Tawi finally quits the hills which overhang it, terminating in a line of cliffs facing the river. It then flows through an open plain.

The Tawi is usually about 92 meters wide and is fordable when the waters are at their lowest, but during the rains it is subject to freshets, when the river rises very suddenly and is transformed into a mighty torrent, about 275 meters in width. Vigne (1842) states that the word "tawi" means a torrent, which aptly describes the river during the rainy season.

The Naoshera Tawi rises in the Rattan Panjal mountains on the road from Thannamandi to Bahramgala, and then turning off to the south-northeast within 0.8 Kms. of the town of Manawar and , 6.5 to 9.5. Kms. beyond, falls into the Chenab near Kuri, a village on the banks of the river.

The Rajouriwali Tawi is a stream of no great depth and contains but little water, excepting when its flow is swollen by rains and the melting of the snows on the Rattan Pir, whence it flows. The river-bed consists of small rocks and rounded stones.

B. *The Jhelum*: The Jhelum originates from the famous spring of Verinag situated at the bottom of a mountain spur in the Pir Panjal range. After flowing through Kashmir Valley, the river enters Pakistan-occupied territory downstream of Baramulla district. The principal tributary of Jhelum in Jammu is Poonch River, having about 110 Kms. of stream-length. It rises on the western slopes of the Panjal range and, as the Sooran, takes a westerly and north-westerly course to the western extremity of Poonch Valley, where it is joined on its right bank by a considerable stream from the direction of Mandi; its course through Poonch Valley is nearly due west, receiving in its passage the waters of Betarh river by its right bank, just west of the town of Poonch. On leaving Poonch Valley, at which point it is joined by the Swan stream from the northwest, it turns to the south and flowing continuously in that direction empties itself into the Jhelum near Tangrot.

The Poonch River in its passage through the Poonch Valley, and as far as the junction of the Mendola river, its stream is broad and comparatively shallow, with, in most places, a moderate current. Between the confluence of the Mendola, and until within a few miles north of Chowmukh, the banks are generally steep and rocky, and the current tempestuous. The confluence of the Poonch River with Jhelum River lies in Pakistan.

C. *The Ravi* : It arises from Bharmaur region of Chamba district in Himachal Pradesh and enters Jammu Province in Basohli tehsil of Kathua district. It is only the right bank of the Ravi which falls within the territory of Jammu & Kashmir State. The principal tributaries of the river are Uj, Basanter, Behani, Khad Kathua and Nalla Bajoi, which have an approximate stream-length of 313 Kms. The confluences of the Uj and the Basanter with the Ravi are located in Pakistan territory.

The majority of the Principal tributaries of the three river systems are perennial and carry crystal clear water throughout the year, except during the monsoons in the catchment areas. Some of the streams have a rapid flow of water with substrata ranging from rocky to sandy. On the basis of the physico-chemical factors and the fauna present in the stream, two categories of streams have been recognized in the entire Province :

(1) Streams with fast current, forming rapids and pools alternately; substrata having large and small stones covered with mucilaginous algal mats; low water temperature and abundance of nymphs of mayfly and stonefly; and *Schizothorax, Nemacheilus* and *Glyptothorax* as the dominant fish species. Such streams have been named as "snow-trout streams", e.g. Darhal, Thannamandi, Rajouriwali Tawi, Betarh, Poonch, Ravi, Pahi, Khad Kathua, Nalla Bajoi and Basanter rivers.

(2) Streams with moderate current, rocky pools at irregular intervals with algal mats; higher water temperature; deficient in the nymphs of mayfly and stone-fly and having *Tor putitora, Garra gotyla* and *Labeo* sps. as the dominant fish species. Such streams have been named as the "mahseer streams", e.g. Tawi, Naddar, Duddar, Jhajjarkotli, Neeladhab, Uj, Anji, Behani and Chenab.

CLIMATE

The location and physiography of the Jammu & Kashmir State is mostly responsible for the varying climatic conditions in its three regions- the Outer Plains and the Outer Hills of Jammu, the Valley of Kashmir and the frontier regions of Ladakh and Kargil. The variations in temperature and rainfall are related to their altitude. Jammu, with an average altitude of about 305 masl., experiences tropical heat and the climatic conditions are just like those of the plains of the Punjab. Climatically the Outer Plain and the Outer Hills of Jammu can be grouped into one region. Its total annual rainfall is about 100 Cms. which gradually increases towards the North. The hot season, as in the Punjab, begins in the middle of March and lasts till the end of June when the summer monsoon rains begin, which last till the end of September.

In the Middle Mountains, which enclose the Valleys of Doda and Kishtwar, the climatic conditions vary from place to place in accordance with their elevations. Kishtwar experiences an extremely rigorous and long winter with a moderate rise in temperature during summer.

The Pir Panjal range, which forms the boundary of the Valley of Kashmir on the South and West, is 80 to 112 Kms. in breadth and has an altitude of 4572 masl. This seems to catch the tail of the south-west monsoons of India and there is, hence, a negligible rainfall in summer. But if unfortunately the monsoon clouds cross the mountain barrier in sufficient quantity, the Valley suffers from serious floods. The warm rain melts the snow on the mountain peaks and floods the Jhelum and its tributaries, causing havoc to the crops, buildings and bridges.

Thus, taking into consideration the temperature and rainfall, the year in the plains and the southern hilly tracts of Jammu Province is divisible into three distinct seasons :

(1) The cold weather season (November to February)

(2) The hot weather season (March to mid-June)

(3) The season of general rains (mid-June to mid-October).

SYSTEMATIC LIST OF FISHES OF JAMMU PROVINCE (J. & K. STATE) INDIA

Family	*Species*	*Distribution in Jammu Province*
I. Notopteridae	I. *Notopterus chitala* (Hamilton)	Tawi
	2. *N. notopterus* (Pallas)	Tawi and Chenab
II. Cyprinidae	3. *Schizopyge curvifrons* Heckel	Suran, Betarh, Poonch and Mandi
	4. *S. longipinnis* Heckel	Suran, Betarh, Poonch and Mandi
	5. *S. progastus* (McClelland)	Poonch
	6. *Schizothorax richardsonii* (Gray)	Tawi, Kud, Mandir Dasala, Binisangh, Katal, Neeru, Thanamandi, Darhal, Rajouri, Budhal, Sunderbani, Betarh, Poonch, Banihal, Ranbir canal, Chenab and Ravi
	7. *Chela (Chela) cachius* (Hamilton)	Choe nallah and Tawi
	8. *C. (Chela) laubuca* (Hamilton)	RS Pora ponds
	9. *Salmostoma bacaila* (Hamilton)	Tawi and Gadigarh fish ponds
	10. *Esomus danricus* (Hamilton)	Tawi, Gadigarh and RSPora fish ponds
	11. *Danio (Danio) devario* (Hamilton)	Tawi, Gadigarh fish ponds, RSPora, Khad Kathua and nalla Bajoi
	12. *D. (Brachydanio) rerio* (Hamilton)	Khad Kathua, nalla Bajoi and Gadigarh fish ponds
	13. *Rasbora (Rasbora) daniconius* (Hamilton)	Khad Kathua, nalla Bajoi and Ravi
	14. *R. (Rasbora) rasbora* (Hamilton)	Gadigarh, Mansar Lake and RSPora fish ponds
	15. *Aspidoparia morar* (Hamilton)	Tawi

(Contd.)

Family	Species	Distribution in Jammu Province
	16. *Amblypharyngodon mola* (Hamilton)	Gadigarh fish ponds
	17. *Barilius barila* (Hamilton)	Dharamsal, Nagrota and Tawi
	18. *B. bendelisis bendelisis* (Hamilton)	Udhampur, Tawi, Rajouri, Sunderbani, Surankote, Poonch and Mandi
	19. *B. bendelisis chedra* (Hamilton)	Udhampur, Tawi, Nagrota, Rajouri, Sunderbani, Betarh, Poonch, Mandi, Naddar, Duddar, Jhajjarkotli, Thannamandi, Basanter, Behani, Reasi, Gadigarh fish ponds and Chenab.
	20. *B. shacra* (Hamilton)	Choe nallah
	21. *B. vagra vagra* (Hamilton)	Tawi, Udhampur, Chineni, Jhajjar, Darhal, Betarh, Uj, Basanter, Chenab, Gadigarh fish ponds
	22. *Puntius amphibius* (Valenciennes)	Poonch
	23. *P. chola* (Hamilton)	Tawi
	24. *P. chrysopterus* (McClelland)	RSPora fish ponds
	25. *P. conchonius* (Hamilton)	Tawi, Gadigarh fish ponds, Mansar Lake and Poonch
	26. *P. sarana sarana* (Hamilton)	Gadigarh fish ponds and Tawi
	27. *P. sophore* (Hamilton)	Tawi, Udhampur, Gadigarh fish ponds and RSPora
	28. *P. ticto ticto* (Hamilton)	Gadigarh and RSPora fish ponds, Udhampur, Tawi, Poonch, Surankote, Muradpur, Nowshera, Sunderbani, Khad Kathua, nalla Bajoi and Chenab
	29. *Osteobrama cotio cotio* (Hamilton)	Gadigarh and RSPora fish ponds
	30. *Labeo dero* (Hamilton)	Udhampur, Tawi, Duddar, Jhajjarkotli, Dharamsal, Khad Kathua, nalla Bajoi and Poonch
	31. *L. bata* (Hamilton)	Tawi
	32. *L. boggut* (Sykes)	Tawi
	33. *L. calbasu* (Hamilton)	Chenab and Tawi
	34. *L. dyocheilus dyocheilus* McClelland	Udhampur, Jhajjarkotli, Tawi and Poonch

(Contd.)

Family	*Species*	*Distribution in Jammu Province*
	35. *L. microphthalmus* Day	Chenab and Tawi
	36. *L. pangusia* (Hamilton)	Tawi
	37. *L. rohita* (Hamilton)	Tawi
	38. *Tor khudree* (Sykes)	Tawi
	39. *T. mosal* (Hamilton)	Jhajjar nallah
	40. *T. tor* (Hamilton)	Chenab and Tawi
	41. *T. putitora* (Hamilton)	Tawi, Naddar, Duddar, Jhajjar-kotli, Pahi, Neeladhab, Rajouri, Poonch, Betarh, Shakhimadan Uj, Behani, Chenab, Birwan Nallah, Nadiali nallah, Kawa daber, Siaran nallah, Junhi nallah, Pullin nallah and Sarande nallah.
	42. *Cirrhinus* (*Cirrhinus*) *mrigala mrigala* (Hamilton)	Tawi and Chenab
	43. *C.* (*Henicorhynchus*) *reba* (Hamilton)	Salhar nallah and Gadigarh fish ponds
	44. *Crossocheilus latius latius* (Hamilton)	Tawi, Darhal, Thannamandi, Betarh, Shakhi- madan, Basan-ter, Khad Kathua, nalla Bajoi.
	45. *C. latius punjabensis* Mukerji	Jhajjar, Ramnagar nalla, Duddar, Tawi, Udhampur, Rajouri, Buddhal, Sunderbani, Poonch and Betarh.
	46. *Garra gotyla gotyla* (Gray)	Tawi, Naddar, Duddar, Jhajjarkotli, Behani, Pahi, Neeladhab, Betarh, Basanter, Khad Kathua, nalla Bajoi, Chenab, Thalli nallah, Sakanwahi Bauli, Devak Nalla, Ramnagar nalla.
	47. *G. lamta* (Hamilton)	Thakarakote
III. Cobitididae	48. *Nemacheilus* (*Acanthocobitis*) *botia aureus* Day	Tawi, Khad Kathua, nalla Bajoi, Birwan nalla and Jhajjar nalla.
	49. *N.* (*Acanthocobitis*) *botia botia* (*Hamilton*)	Gadigarh ponds, Tawi, Darhal, Thannamandi, Betarh, Basanter, Behani, Chenab, Udhampur and Poonch.
	50. *N.* (*Nemacheilus*) *corica* (Hamilton).	Tawi, Nagrota nallah and Sarande nallah.
	51. *N.* (*Schistura*) *kangrae* Menon	Tawi, Naddar, Jhajjarkotli, Anji, Neeladhab, Darhal, Betarh, Chenab, Shakhimadan, Poonch and Ravi.

(Contd.)

Family	*Species*	*Distribution in Jammu Province*
	52. *N.* (*Schistura*) *montanus* (McClelland)	Loran (Poonch)
	53. *N.* (*Schistura*) *prashari prashari Hora*	Bali nallah, Kud nallah, Mandir Dasala nalla, Binisangh nalla, Samroli nalla, Arsa nalla, Junhi nalla, Jhajjar nalla, Nagrota nallah and Tawi,
	54. *N.* (*Schistura*) *rupecola rupecola* (McClelland).	Tawi, Chenah and Poonch.
	55. *N.* (*Schistura*)*sinuatus* Day	Betarh nallah
	56. *N.* (*Schistura*) *striatus* Day	Mandi (Poonch)
	57. *N.* (*Schistura*) *vittatus* (Heckel)	Sooran torrent
	58. *N.* (*Triplophysa*) *gracilis* Day	Poonch, Surankote, Rajouri, Buddhal and Sunderbani
	59. *N.* (*Triplophysa*) *kashmirensis Hora*	Rajouri, Budhal, Sunderbani, Surankote and Poonch.
	60. *N.* (*Triplophysa*) *marmoratus* (Heckel).	Sooran torrent
	61. *Botia* (*Botia*) *almorhae* Gray	Betarh and Poonch
	62. *B.* (*Botia*) *birdi* Chaudhuri	Jhajjar nallah
	63. *B.* (*Botia*) *rostrata* Gunther	Eak nallah and Tawi.
	64. *Lepidocephalus* (*Lepidocephalichthys*) *guntea* (Hamilton)	Nadiali nalla, Birwan nalla, Siaran nalla, Dhanodi nalla and Choe nalla.
IV. Bagridae	65. *Aorichthys seenghala* (Sykes)	Tawi and Gadigarh fish ponds
	66. *Mystus bleekeri* (Day)	Gadigarh and RSPora fish ponds, Bhaderwah
	67. *M. vittatus* (*Bloch*)	Gadigarh and RSPora fish ponds.
V. Siluridae	68. *Ompok bimaculatus* (Bloch)	Gadigarh and RSPora fish ponds.
	69. *Wallago attu* (Schneider)	Gadigarh and RSPora fish ponds and lower reaches of river Tawi.
VI. Schilbeidae	70. *Clupisoma garua* (Hamilton)	Poonch river.
	71. *Eutropiichthys vacha* (Hamilton)	Lower reaches of Tawi, River Chenab.
VII. Amblyci-pitidae	72. *Amblyceps mangois* (Hamilton)	Khad Kathua, nalla Bajoi, Jhajjar nallah and Tawi.
VIII. Sisoridae	73. *Bagarius bagarius* (Hamilton)	Lower reaches of Chenab, Tawi.

(Contd.)

Family	*Species*	*Distribution in Jammu Province*
	74. *Glyptosternum reticulatum* McClelland.	Rajouri and Poonch.
	75. *Glyptothorax cavia* (Hamilton)	Betarh and Poonch.
	76. *G. conirostre conirostre* (Steindachner).	Poonch and Sooran torrent.
	77. *G. kashmirensis* Hora	Rajouri, Poonch and Betarh.
	78. *G. pectinoperus* (McClelland)	Pahi, Neeladhab, Thannamandi, Betarh, Khad Kathua, nalla Bajoi, Poonch, Chenab, Jhallas nalla, Sooran, Devak nalla, Dhanodi nalla, Kud nalla,Junhi nalla, Jhajjar nalla, Puthi Khad nalla, Tawi and Bhaderwah.
	79. *G. telchitta telchitta* (Hamilton)	Salahar.
IX. Hetero-pneustidae	80. *Heteropneustes fossilis* (Bloch)	Gadigarh and RSPora fish ponds, Tawi and Bhaderwah.
X. Belonidae	81. *Xenentodon cancila* (Hamilton)	Gadigarh fish ponds, Tawi, Meander, Khad Kathua and Nalla Bajoi.
XI. Channidae	82. *Channa marulius* Hamilton	Choe nalla, Tawi.
	83. *C. orientalis* (Schneider)	Nadiali Nalla, Birwan nalla, Thalli Nalla, Devak nalla, Dhanodi nalla, Siaran nalla, Garhi nalla, Jhajjar nalla, Choe nalla, Sarande nalla, Drungli nalla, Gadigarh, Mansar Lake, RSPora fish ponds and Uj.
	84. *C. punctatus* (Bloch)	Drungli, Choe nalla and Tawi.
	85. *C. striatus* (Bloch)	Poonch and lower reaches of Tawi.
XII. Chandidae	86. *Chanda nama* (Hamilton)	Lower reaches of Tawi.
XIII. Nandidae	87. *Badis badis* (Hamilton)	Gadigarh fish ponds and Choe nalla.
	88. *Nandus nandus* (Hamilton)	Gadigarh fish ponds.
XIV. Gobiidae	89. *Glossogobius giuris giuris* (Hamilton)	Gadigarh fish ponds
XV. Belontidae	90. *Colissa fasciatus* (Schneider)	Gadigarh fish ponds

(Contd.)

Family	Species	Distribution in Jammu Province
XVI. Mastacembelidae	91. *Macrognathus aculeatus* (Bloch)	Salahar
	92. *Mastacembelus armatus armatus* (Lacepede).	Gadigarh and RSPora fish ponds, Shakhimadan, Behani, Khad Kathua, nalla Bajoi, Nadiali nalla, Dhanodi nalla Tawi, Jhajjar, Junhi nalla, Betarh and Poonch.
	93. *M. pancalus* (Hamilton)	Gadigarh and RSPora fish ponds and Tawi.

ICHTHYOGEOGRAPHY

Palaeoichthyogeography

From time to time widely divergent views have been put forth to explain the occurrence and assemblance of freshwater ichthyofauna of India in relation to adjacent countries (Jhingran 1985) : Gunther (1880) observed a great affinity between the fish faunae of the Indian and African regions and showed that one or more representatives of the seventeen out of twenty six families or groups found in the Indian region exist in Africa; and many African species are not even generically different from those of India. Day (1885), on the other hand, found eighty seven genera of freshwater teleosts in India (including within it Burma and Sri Lanka), of which only fourteen were represented in Africa. He also observed that of the 369 species which belong to these genera, the distribution of only fourteen species extended to Africa.

Blanford (1901) supported Day's view and concluded that the elements in Indo-Malayan fauna are heavily represented in the Eastern Himalayas, diminishing gradually westwards and ceasing altogether in Kashmir and further West.

Hora (1937) also supported Blanford's view and advocated a Southeast-Asian origin of the freshwater fish fauna of India, emanating most porbably from Yunnan (in China). According to Hora's view this ichthyofauna spread westwards in succcessive waves of migration, first to India and later on to Africa, at a time when the two landmasses were connected. Hora (1944) critically studied the 41 Indian fish genera and concluded that not only had the Indian fish fauna a marked Malayan affinity but was related to the faunae of Thailand, Southern China and Cochin-China.

After a detailed analysis of 89 genera of Indian freshwater fishes Menon (1955) reiterated Hora's hypothesis that the ichthyofauna of India was related to Malayan ichthyofauna (of the East). An examination of the Indian endemic genera also shows a close affinity with forms in countries towards the East of India, viz. Burma, Malaya, Sumatra and Java.

During the early Eocene, *Lepisosteus* Lacepede and certain teleosts belonging to the families Osteoglossidae, Cyprinidae, Anabantidae and Nandidae migrated from the North to the South into India. Their fossils have been found in Deothan, Kheri, Takli and Paharsingha in Madhya Pradesh.

During the Pliocene there occurred a dispersal of fishes inhabiting the marshes, as is evidenced from the fossil relics of genera such as *Clarias* Scopoli, *Heterobranchus* Geoffr., *Channa* Scopoli, *Rita* Bleeker, *Bagarius* Bleeker, *Silurus* Linn., *Chrysichthys* Bleeker, *Mystus* Scopoli and various Ophiocephalids found in the Siwalik ranges of India (Hora and Menon 1953). Fishes typical of the smaller torrential streams are found to have followed a more-or-less parallel history of emergence in the early Pleistocene. The typical Himalayan ichthyofauna such as *Nemacheilus* van Hasselt, *Botia* Gray, *Garra* Hamilton, *Tor* Gray, *Schizothorax* Heckel etc., undoubtedly originated somewhere in the Palaearctic South China and dispersed into regions lying along the Alpine - Himalayan system and its southern loops ramifying later into Europe, Western Asia and other associated mountains (Hora 1951).

Most of the present day fish fauna of India has descended from the Pleistocene migrants; while the migration along the Himalayas westwards commenced in the Pliocene and continued uninterrupted till the late Pleistocene, when the reversal of the westward-flowing drainage at the foot of the Himalayas had occurred. The migration southwards to the Peninsula had occurred entirely during the Pleistocene.

Neoichthyogeography

The ichthyofaunal interest of Jammu Province lies in the unique phenomenon of "East meets West", for some regions of the Province (such as Poonch Valley) exhibit a rare combination of Palaearctic fish species with fishes of Indian origin (Das & Nath 1971, Nath 1992). The origin of fishes of Jammu Province is from the Indo-Gangetic region (and a few species from Kashmir region), to which region there was migration from East to West from Southeast Asia. Evidence exists that the fishes of Jammu region have probably migrated to their present habitat only in the late Pleistocene (Das & Nath 1971, Nath 1992). Moreover, as many as 56 species of freshwater teleosts of the River Kosi drainage system are represented in the ichthyofauna of Jammu region (Khan & Kamal1979, Nath 1986). Several freshwater teleosts of Iran also find representation in the ichthyofauna of Jammu region (Berg 1949, Coad 1979, Nath 1986).

A comparative study of the ichthyofauna of Kashmir Valley and Jammu Province reveals that the nearly complete isolation of Kashmir region from Jammu region by the Pir Panjal range mountain barrier has excluded many of the fish species of Kashmir region from Jammu region and vice versa. The low endemicity of the fish fauna in the river systems of Jammu region in contrast to the much older river systems of Kashmir indicates that probably a geologically-sufficient time-lapse has not occurred for the fishes in Jammu region to evolve into new species by isolation. That a few fishes (*e.g. Schizothorax richardsonii, Schizopyge progastus, Labeo dero, Nemacheilus kashmirensis*, *Glyptothorax kashmirensis* and *Glyptosternum reticulatum* should have established themselves across lofty mountain barriers in Jammu region is yet another evidence of the great climatic and ecological tolerance exhibited by these teleosts, which probably established together in both regions in the late Pleistocene.

The total absence of the representatives of the families Bagridae, Schilbeidae, Heteropneustidae, Amblycipitidae, Mastacembelidae, Belonidae, Gobiidae, Belontidae, Channidae and Chandidae of Jammu region from Kashmir Valley suggests that the high Pir Panjal range was already established when these fishes of Indian region migrated towards Kashmir but were prevented from reaching Kashmir Valley by the Pir Panjal range.

The occurrence of *Schizothorax richardsonii, Schizopyge progastus, Garra gotyla gotyla, Tor putitora, Labeo dero, Nemacheilus* sps., *Glyptothorax* sps. and *Glyptosternum reticulatum* in the fast streams of Poonch Valley in Jammu Province is not surprising, as all these teleosts possess adequate adaptive modifications to allow them to live in torrential streams. Prolonged evolution under severe mountain-terrain conditions resulted in the development of special adaptive mechanisms and subsequent modifications (e.g. reduction of scales, number of pharyngeal teeth and the number of barbels, reduced or depressed bodies, etc.) in the torrential stream fishes.

The fish species of the River Kosi drainage system which are represented in the ichthyofauna of Jammu Province are : *Notopterus notopterus, Chela laubuca, Salmostoma bacaila, Barilius barila, B.bendelisis, B. shacra, B.vagra, Danio devario, Rasbora daniconius, Amblypharyngodon mola, Aspidoparia morar, Puntius chola, P. conchonius, P. sarana, P. sophore, P. ticto, Tor putitora, Cirrhinus mrigala, C. reba, Crossocheilus latius latius, Garra gotyla gotyla, G. lamta, Labeo bata, L. calbasu, L. dero, L. pangusia, L. rohita, Osteobrama cotio, Schizothorax richardsonii, Botia dayi, Lepidocephalus guntea, Nemacheilus botia, N. rupecola, Ompok bimaculatus, Wallago attu, Mystus bleekeri, M. vittatus, Aorichthys seenghala, Amblyceps mangois, Bagarius bagarius, Glyptothorax cavia, G. telchitta, Clupisoma garua, Eutropiichthys vacha, Heteropneustes fossilis, Xenentodon cancila, Channa orientalis, C. punctatus, C. striatus, C. marulius, Chanda nama, Colisa fasciatus, Glossogobius giuris, Macrognathus aculeatus, Mastacembelus armatus* and *M. pancalus.*

Aspidoparia morar, Crossocheilus latius diplocheilus, Labeo dero, Nemacheilus prashari, Heteropneustes fossilis and *Channa orientalis* are the freshwater teleosts of Jammu Province which are also represented in the ichthyofauna of Iran (Berg 1949, Coad 1979).

The ichthyofauna of Jammu Province is dominated by the representatives of the families Cyprinidae (44 species) and Cobitididae (17 species). *Glyptothorax cavia, G. pectinopterus,* several species of *Nemacheilus* and *Tor putitora* are the most highly evolved torrential fishes of Jammu Province, while *Garra gotyla gotyla, G. lamta, Nemacheilus* sps., etc. are hill-stream forms and generally inhabit streams with a rocky bottom.

The ichthyofaunal interest of Poonch region of Jammu Province lies in the unique phenomenon of "East meets West", for this is the only region which exhibits a rare combination of Palaearctic fish species with Indo-Malayan fishes. This is, in fact, true not only of fishes but also of several other vertebrate and invertebrate groups such as mammals, herpetiles, molluscs, crustaceans, etc. (Nath 1980, 1984, 1985, 1986).

A comparison of the ichthyofaunal elements of Kashmir Valley with those of Poonch Valley of Jammu Province reveals the following salient features :

1. *Schizopyge progastus, S. curvifrons, S. longipinnis, Schizothorax richardsonii, Puntius conchonius, P. sophore, Labeo dero, L. dyocheilus, Tor putitora, Crossocheilus latius punjabensis, Garra gotyla gotyla, Botia almorhae, B. birdi, Nemacheilus marmoratus, N. vittatus, N. rupecola, N. kashmirensis, N. gracilis, Glyptothorax kashmirensis* and *Glyptosternum reticulatum* are the only species which are common to both regions.
2. The relatively-low endemicity of the fish fauna of the river systems of Poonch region is fairly obvious. It is probable that a geologically sufficient time-lapse has not occurred for the fish species in Poonch Valley to evolve into new endemic species, which is in sharp contrast to the much older river systems of Kashmir Valley.
3. Das (1966) elucidated that there are at least ten Palaearctic fish genera thriving in the freshwaters of Kashmir Valley. On the contrary, the Palaearctic element in the ichthyofauna of Poonch Valley is represented only by *Schizothorax richardsonii, Schiz-*

opyge progastus and *Glyptosternum reticulatum*, most of the other species being of Indo-Malayan origin.

4. *Schizothorax* sps. and *Glyptothorax telchitta* are found along other regions of the Himalayas (e.g. Kumaon Range) and are torrential stream fishes that were carried along hill-streams to the Indian region as well.
5. *Schizopyge* is possibly the oldest fish to have established from the Palaearctic region into Kashmir, with more than six endemic species in Kashmir Valley. Of these a few have been carried to the Indian Himalayas and Poonch Valley (*Schizopyge progastus*,*S. longipinnis*, *S. curvifrons* and *Schizothorax richardsonii*). These species appear to have evolved by isolation in Kashmir from the Palaearctic species *Schizothorax intermedius*, still found in Russia, S. China and Central Asia, but only three species have been able to reach Poonch Valley and adjacent areas of Jammu Province.

According to Menon (1955), the fish genera endemic to and exclusive for India are : *Parasilorhynchus* Hora, *Lepidopygopsis* Raj, *Balitora* Gray, *Bhavania* Gray, *Travancoria* Hora, *Aborichthys* Chaudhuri, *Jerdonia* Day, *Noemachilichthys* Day, *Somileptes* Swainson, *Horabagrus* Jayaram, *Hara* Blyth, *Myersglanis* Hora & Silas, *Sisor* Hamilton, *Ctenops* McClelland, *Ailia* Gray, *Neotropius* Kulkarni, *Proeutropiichthys* Hora, *Silonia* Swainson, *Horaglanis* Menon and *Amphipnous* Muller. It is surprising that all these endemic fish genera of India are totally absent from Jammu Province, as has been confirmed by Jhingran (1974) and Jayaram (1981).

The ichthyofauna of Jammu Province can be further classified as under :

A. Widely distributed species :

I. Species found in Sri Lanka, India, Burma and Farther East :
Rasbora daniconius, *Puntius sarana sarana*, *P. ticto ticto*, *Nemacheilus botia*, *Ompok bimaculatus*, *Wallago attu*, *Mystus vittatus*, *Heteropneustes fossilis*, *Xenentodon cancila*, *Channa orientalis*, *C. marulius*, *C. striatus* and *Mastacembelus armatus*.

II. Species found in India, Burma and Farther East but absent from Sri Lanka :
Notopterus notopterus, *Rasbora rasbora*, *Puntius chola*, *P. sophore*, *Aorichthys seenghala*, *Bagarius bagarius*,*Channa punctatus* and *Glossogobius giuris*.

III. Species confined to India and Burma only :
Chela cachius, *C.laubuca*, *Salmostoma bacaila*, *Barilius bendelisis chedra*, *Amblypharyngodon mola*, *Cirrhinus mrigala*, *Chanda nama*, *Badis badis* and *Labeo calbasu*.

IV. Species confined to India only :
Cirrhinus reba and *Mastacembelus pancalus*.

B. Species of Northern India :

I. Species found in Northern India, Burma and Further East :
Notopterus chitala, *Labeo dyocheilus* and *Eutropiichthys vacha*.

II. Species found in Northern India and Burma :
Barilius barila, *Danio* (*Brachydanio*) *rerio*, *Esomus danricus* (also found in Sri Lanka), *Aspidoparia morar*, *Labeo rohita*, *Mystus bleekeri*, *Amblyceps mangois*, *Clupisoma garua* and *Colisa fasciatus*.

III. Species found in Northern India only :
Barilius vagra, *B. shacra*, *Danio* (*Danio*) *devario*, *Aspidoparia morar*, *Puntius conchonius*, *P. chola*, *Tor putitora*, *Labeo bata*, *L. pangusia*, *Osteobrama cotio cotio*, *Crossocheilus latius latius*, *Lepidocephalus guntea* and *Macrognathus aculeatus*.

C. Species of the Himalayas :

I. Species of Jammu fishes found in all the four sections of the Himalayas (Assam, Nepal, Kumaon and Punjab) :
Nemacheilus corica (Tawi and its tributaries)

II. Species of Jammu fishes found in Assam Himalayan and Kumaon Himalayas only:
Glyptothorax telchitta (Salahar area)

III. Species of Jammu fishes endemic in Assam Himalayas :
Schizopyge progastus (Poonch streams).

Moreover, out of the 31 species of freshwater teleosts recorded from River Alaknanda (Badola and Singh 1981) as many as 17 species also occur in Jammu Province, viz. *Schizothorax richardsonii, Tor putitora, Labeo dero, L. dyocheilus, Barilius bendelisis, B. barila, B. vagra, Garra gotyla gotyla, G. lamta,Crossocheilus latius latius, Nemacheilus rupecola, N. montanus, Glyptothorax pectinopterus, G. cavia, G. conirostre, G. telchitta* and *Mastacembelus armatus.*

In the opinion of the present author, the ichthyofauna of Jammu Province is mostly composed of Himalayan elements, as is indicated by the preponderance of Cyprinid and Cobitidid representatives. Some fishes of Palaearctic region have been able to reach Jammu Province before the rising of the present mountain barrier (Pir Panjal) isolating Jammu from Kashmir region. It is also evident that the mountain barrier between the two regions has prevented the entry of the majority of endemic fish species of Kashmir region into Jammu Province. At the same time it is significant that all the endemic and exclusive fish-species of Indian region have failed to reach Jammu Province (via Punjab for instance) in spite of the contiguity of the two regions and the absence of any formidable mountain barriers between them. The representation of several fish species of Eastern India in the ichthyofauna of Jammu Province as well as the countries situated to the West of the region (such as Afghanistan, Iran and adjacent countries) indicates that the Jammu Province of Jammu and Kashmir State may have acted as an "aqueduct" for the passage of these fishes from East to West. This is true not only of fishes but also of a large number of other animals such as crustaceans, molluscs, herpetiles, birds and mammals (Nath 1980, 1984, 1985, 1986, 1988, 1989, 1993).

The occurrence of a large number of fishes of Eastern India in the freshwaters of Jammu Province seems to indicate that in the remote past the drainage systems of the whole of Northern India must have been confluent with each other, having got separated only much later, after the fishes of the eastern region had an opportunity to migrate to and establish themselves in the river systems of the West.

A detailed taxonomic study of the fish specimens obtained from Jammu Province reveals that many of them differ from the holotypes in several details, e.g. *Barilius bendelisis bendelisis, B. bendelisis chedra, B. barila, B. shacra, Schizothorax richardsonii, Garra gotyla gotyla, Puntius ticto ticto, P. sophore, Amblypharyngodon mola, Labeo dero, L. dyocheilus,Crossocheilus latius, Nemacheilus botia, N. prashari, Amblyceps mangois, Glyptothorax pectinopterus* and *Channa orientalis*. While many of these differences are due to racial variations, some of them indicate that the fishes of Jammu region have already started to evolve into new species / varieties inspite of the geologically short time-lapse which has occurred since the isolation of Jammu region from Kashmir region.

Lastly, the occurrence of teleosts such as *Tor khudree, Nemacheilus sinuatus* and *N. striatus* in Jammu Province is yet another salient feature of the ichthyofauna of this region, as the said species were previously known from Western and Southern India only. Apparently these species have a much wider range of distribution than was previously known and extends to Northern India as well.

REFERENCES

Badola, S.P. and Singh, H.R. 1981. Fish and fisheries of the river Alaknanda. *Proc. natn. Acad.Sci. India, 51 B* (I): 133 - 142.

Berg, L.S. 1948-49. *Freshwater Fishes of U.S.S.R. and adjacent Countries.* Leningrad, 3.

Berg, L.S. 1949. *Freshwater Fishes of Iran and Adjacent Countries. Trudy Zool. Inst. Akad. Nauk SSSR, 8* : 783 - 858.

Blanford, W.T. 1901.The distribution of vertebrate animals in India, Ceylon and Burma. *Philos. Trans. Roy. Soc., 194 B* : 335-436.

Bleeker, P. Von 1853. Nalezingen op de ichthyologische fauna van Bengalen en Hindostan. *Verh. Batavia Genoot., 25*: I-164.

Coad, W. Brian 1979. A provisional annotated checklist of the freshwater fishes of Iran. *J. Bombay nat. Hist. Soc., 76* (I) : 86-105.

Das, S.M. 1965. A revision of the fish species inhabiting Kashmir Province. *Kashmir Sci,.* 2 (1-2) : 13-19.

Das, S.M. 1966. Palaearctic elements in the fauna of Kashmir. *Nature, 212* : I-6.

Das, S.M. and Nath, Surendra 1965. The ichthyofauna of Poonch Valley (Jammu and Kashmir State). *Kashmir Sci.,* 2 (I-2) : 147-155.

Das, S.M. and Nath, Surendra 1966. The ichthyofauna of Jammu Province, J. & K. State, *Ibid., 3* (I-2) : 65-78.

Das, S.M. and Nath, Surendra 1971. A revision of fishes from Jammu Province India. *Ibid., 8* (I-2) : I-22.

Das, S.M. and Subla, B.A. 1963.The ichthyofauna of Kashmir, Part I : History, topography, origin, ecology and general distribution. *Ichthyologica, 2*(I-2) : 87-106.

Das, S.M. and Subla, B.A. 1964. The ichthyofauna of Kashmir, Part II : The speciation of Kashmir fishes with two new records of species. *Ibid.,3*(I-2) : 57-62.

Das, S.M. and Subla, B.A. 1970. The Pamir-Kashmir theory of origin, evolution and distribution of Kashmir fishes, with their general ecology. *Kashmir Sci., 7* (I-2) : I -15.

Day, F. 1885. Relationships of the Indian and African freshwater fish fauna. *J. Linn. Soc.* (*Zool.*), *18* : 308-317.

Day, F.1889. The fauna of British India including Ceylon and Burma. Fishes. London, Taylor & Francis.

De Witt, H. 1960. A contribution to the ichthyology of Nepal. *Stanford Ichthyol. Bull.*, No: *4*,7 : 63-88.

Hora, S.L. 1937. Geographical distribution of Indian freshwater fishes and its bearing on the probable land connection between India and the adjacent countries. *Curr. Sci., 7* : 351-356.

Hora, S.L.1937.b. Comparison of the fish fauna of the northern and southern faces of the Great Himalayan range. *Rec. Indian Mus., 39* : 24I- 250.

Hora, S.L.1944. On the Malayan affinities of the freshwater fish fauna of Peninsular India and its bearing on the probable age of the Garo-Rajmahal gap. *Proc. natn. Inst. Sci. India,* I0 (4) : 423-29.

Hora, S.L. 1951. Fish geography of India. *J.Zool. Soc. India, 3* (1) : 187-197.

Hora, S.L. 1951 b. Some observations on the palaeogeography of the Garo-Rajmahal gap as evidenced by the distribution of the Malayan fauna and flora to Peninsular India. *Proc. natn. Inst. Sci. India, 17* : 437-444.

Hora, S.L. 1953. Dispersal of fishes from Central Asia. *Rec. Indian Mus. 41*: 36.

Hora, S.L.1955. Place of Kashmir in the Fish Geography of India. *Everyday Sci., 3*(1) : 36-45.

Hora, S.L., Malik,G.M. and Khajuria, H. 1955. Some interesting features of the aquatic fauna of the Kashmir Valley. *J. Bombay nat. Hist. Soc., 53* (I) : I40-I43.

Hora, S.L.and Menon, A.G.K. 1953. Distribution of Indian fishes of the past and their bearing on the geography of India (2) : The extinct freshwater teleostean fishes of India. *Everyday Sci., 2* : 105-113.

Hora, S.L. and Silas, E.G. 1952. Evolution and distribution of Glyptosternoid fishes of the family Sisoridae (Order: Siluroidea). *Proc. natn. Inst. Sci., India, 18*(4) : 309-322.

Jayaram, K.C. 1974. Ecology and distribution of freshwater fishes, amphibia and reptiles. In: Mani, MS. (Ed.) I974: *Ecology and Biogeography of India.* Dr. W.Junk, Hague, pp. 517-584.

Jayaram, K.C. 1981. *The freshwater fishes of India, Pakistan Bangladesh, Burma and Sri Lanka - A Handbook.* The Zoological Survey of India, Calcutta.

Jerdon, T.C. 1849. On the freshwater fishes of southern India. *Madras J. Lit. Sci., 15* : 302 -346.

Jhingran, V.G. 1975. *Fish and Fisheries of India.* Hindustan Publishing Corpn. Delhi.

Joshi, C.B., Sehgal ,K.L. and Sunder, Shyam 1978. Observations on the fishery resources of the hill-streams of Jammu Province with special reference to Mahseer and other commercially important species. *Indian J. Fish., 25* (1-2) : 197-206.

Jyoti, M.K. and Sehgal, H.S. 1982. Aquatic resources of J.& K. State and their fisheries potential. *Proc. 69th Indian Sci. Congr.*, III : 86.

Khan, H.A. and Kamal, M.Y. 1979. On a collection of fish from River Kosi (Bihar). *J. Bombay nat. Hist. Soc.*, *76*(3): 530-534.

Kottelet, M. 1984. Revision of the Indonesian and Malaysian loaches of the subfamily Noemacheilinae (Osteichthyes : Cypriniformes). *Japan. J. Ichthyol.*, *31* (3) : 225-260.

Kottelet, M. 1987. Nomenclatural status of the fish names created by J.C. van Hasselt (I823) and of some cobitoid genera. *Ibid.*, *33*(4) : 368-375.

Menon, A.G.K. 1952. The external relations of the Indian freshwater fishes with special reference to the countries bordering on the Indian Ocean. *J.Asiat. Soc.*, *2I* (1) : 3I-38.

Nath, Surendra 1969. On some new records of fish from Jammu Province (J.&K. State). *Proc. Natn. Acad. Sci. India*, Annual No : 1969 : 131.

Nath, Surendra 1972. On the occurrence of *Bubo bubo turcomanus* (Eversmann), (Strigiformes: Strigidae) in Poonch Valley (J.&K. State), India. *Proc. 59th Indian Sci. Congr.*, III : 452.

Nath, Surendra 1973. The Ichthyofauna of Jammu and Kashmir State (India), Part I : Fishes of Jammu (Tawi), with some new records. *Proc. 60th. Indian Sci. Congr.*, III : 559.

Nath, Surendra 1973 b. The Ichthyofauna of Jammu and Kashmir State (India), Part II : On some new records of freshwater teleosts from Poonch Valley (J.& K. State). *Proc. Natn. Acad, Sci. India*, Annual No : 1973 : 118.

Nath, Surendra 1974. Notes on the entomofauna of Jammu and Kashmir State (India), Part I : On the occurrence of *Cybister tripunctatus* Oliv. (Coleoptera ; Adephaga ; Dytiscidae) and *Hydrophilus piceus* Linn. (Polyphaga ; Hydrophilidae) in Kashmir Valley. *Ibid.*, Annual No : 1974.

Nath, Surendra (1980). Extension of range of *Tripos cancriformis* (Bosc) (Branchiopoda; Notostraca; Apodidae) to Poonch Valley (Jammu & Kashmir State). *J. Bombay Nat. Hist. Soc.*, *76* (3) : 543-544.

Nath, Surendra 1981. On the extension of range of two freshwater catfishes, *Glyptothhorax conirostre conirostre* (Steind.) (Sisoridae) and *Clupisoma garua* (Ham.) (Schilbeidae), to Poonch Valley (Jammu and Kashmir), India. *Ibid.*, *78* (1) : 178-179.

Nath, Surendra 1982. Extension of range of the putitor mahseer *Tor putitora* (Ham.) (Cypriniformes; Cyprinidae; Barbinae) to Poonch Valley (Jammu and Kashmir). *Ibid.*, *79* (2) : 430-431.

Nath, Surendra 1983. Extension of range of the river crab - *Potamon* (*Potamon*) *atkinsonianum* Woodmason (Brachyura; Cyclometopa; Potamonidae) to Poonch Valley (Jammu and Kashmir State). *Ibid.*, *79*(3) : 705.

Nath, Surendra 1984. On the extension of range of a freshwater catfish - *Glyptothorax lineatus* (Day) (Siluriformes; Sisoridae) - to Poonch Valley (Jammu and Kashmir State). *Ibid.*, *80* (3) : 651-652.

Nath, Surendra 1985. On the extension of range of the Vespertilionid bat- *Pipistrellus paterculus* (Thomas) - to Poonch Valley (Jammu and Kashmir State) . *Ibid.*, *82* (2) :396.

Nath, Surendra 1986. Observations on the general ecology of fishes of Poonch Valley (Jammu and Kashmir State) India. *Livestock Adviser*, *11* (12) : 31 - 36.

Nath, Surendra 1986 b. On the occurrence of the Amphipod - *Gammarus pulex* (Limn.) - in Kashmir Valley, with remarks on the ecology of the species. *J. Zool. Soc. India*, *38* (I&2) : 115-116.

Nath, Surendra (1986c). A Check-list of fishes of Jammu and Kashmir State (India), with remarks on the ichthyogeography of the State. *Ibid.*, *38* (1&2) : 83-98.

Nath, Surendra (1987). On a collection of Bats of the sub-order Micro-Chiroptera from Poonch Valley and adjacent areas of Jammu and Kashmir State (India), with some new records. *Intrascience*, *2* (3) : 96-98.

Nath, Surendra (1992). Ichthyofauna of Jammu and Kashmir State (India), with remarks on the ichthyogeography of the region. *Rec. Adv. Fish. Ecol. Limn. Eco-Conser.*, *2* : 1 -32.

Nath, Surendra 1993. A checklist of Mammals of Jammu and Kashmir State (India) , with remarks on the Mammal geography of the State. *Livestock Adv.*, *18*(5) : 6-13.

Silas, E.G. 1960. Fishes from Kashmir Valley. *J. Bombay Nat. Hist. Soc.*, *57*(1) : 66-77.

Talwar, P.K. (1978). On the fishes collected by the Ladakh expedition 1976. *Ibid.*, *74* (3) : 501-505.

Tilak, R. (1967). The fishes of river Tawi and its tributaries (Jammu State) with some notes on ecology. *Rec. Zool. Surv. India*, *65* : 183-232.

Rec. Adv. Fish Ecol.Limn. Eco-Conserv.
III : 121-123

14

Nanda Devi Sanctuary—Now a National Park

S.M. DAS

The recent decision of Govt. of India/WWF/Govt. of U.P. to convert Nanda Devi Sanctuary into a National Park is timely and welcome, as it is the highest altitude National Park in India as well as the Himalaya. This Sanctuary is the only wildlife conservation area at the extreme heights of 12,000 to 14,000 feet in Garhwal Himalaya, where the rare endangered wildlife such as Bharal (blue sheep), Musk deer, Snow- leopard, Tibetan Gazelle, Tibetan sheep, Brown bear, Flying squirrel, Snow partridge, Snow cock, Snow pigeon, Chakor, and Monal pheasant, were once abundant.

Nanda Devi base ring is a natural wildlife sanctuary discovered in 1934 by Eric Shipton and H.W.Talman, two mountaineers who attempted the Nanda peak and first found the approach route to the Inner Sanctuary. Since then at least 20 expeditions have passed through the Inner Nanda Devi Sanctuary to attempt the peak top. Many more tourists and foreign visitors have camped in the Outer Sanctuary. The Nainital team reached the outer Sanctuary (12,000 feet) in 1974 and found the remains of trees, cooking fires, tins, bottles, cartons and other man-produced litter scattered on the hill-sides, despoiling the ecology of the area. Some Bharal skulls without horns were also found, sure evidence of poaching and killing.

The Indian Mountaineering Foundation (I.M.F, with Dr. H.C. Sarin as President) prepared a list of do's and don'ts for mountaineers in 1976, for ecological conservation of Nanda Devi Sanctuary. But little attention was paid to these directives, with the result that shooting and poaching continued unabated. Out of a rough count of 800 Bharal in the Sanctuary hardly 300 remain today.

Next the World Wildlife Fund (WWF) sent an expert team to assess the extent of damage and suggest measures to prevent human depredation in the area. The team recommended to the Food & Agriculture Ministry as well as the U.P. Government that even the Inner Sanctuary is being extensively poached by hill village fire-arm holders as well as trekkers and mountaineers for musk (5,000/= a pod), hide, horn and hair-wool. Shepherds also collect, during their short stay in the Inner Sanctuary (15 days to One month), aromatic and medicinal plants of high value. The team also reported signs of deforestation (as in the Outer Sanctuary) and fires and litter to have assumed alarming proportions.

Although the I.M.F. was afraid that by converting Nanda Devi Sanctuary into a National Park, mountaineering and adventure will receive a setback, the recent scientific expedition by Dr. Lamba of Z.S.I. and his team (1980-82), who has checked the pros and cons, has shown that without strict conservation this rare endangered wildlife pocket may disappear altogether. The newly converted National Park is feasible since there is no need to patrol the entire area, as the Inner Sanctuary is guarded by a vast ring of high mountains 70 KM in circumference on all sides. A narrow pass, the Rishi Ganga opening, is the only passage to the Inner Sanctuary which can be easily guarded and all fire-arms deposited with the guard at Lata Village, before entering the Inner Sanctuary. Thus small parties of mountaineers and trekkers can easily be allowed to pass through into the Park with cameras, food, bedding and tents only.

The route starts from Joshimath through Tapovan to Lata (last Motor link); next lies 8 days tough march along steep precipitous rocks, along Rishi Ganga gorge, with rope—climbing necessitated in some places. The route then passes through Lata Kharak, Dharausi, Deodi, Ramni, Bhugarh, Patal Khani and finally Sarso Patal at 14,000 feet (usual base camp). Small parties of trekkers and mountaineers may visit this marvellous Inner Sanctuary in the summer and enjoy primeval nature in high Himalayas, still maintaining the ancestral biological gene pool. Wildlife conservation literature should be distributed to visitors and tourists entering the Valley.

The Nanda Devi National Park lies in the cold rocky and snowy belt of the Himalaya. Few favourable spots exist for permanent settlements, except for a few huts of Bakarwals in the outer Sanctuary. The vegetation consists of small Alpine meadows, Alpine scrub, dwarf *Rhododendron*; and some short season flowers such as *Gentiana, Primula, Iris, Saxifraga*, and *Geranium*. Pure stands of *Betula utilis*, Birches and Junipers are found on the slopes up to 9,000 feet. No Orchids or Brahmak (*Soussouria*) (still found in Garhwal) were found even in the Outer Sanctuary.

Besides wildlife conservation literature for visitors, they should be given seeds and saplings at Lata to plant them on the bare deforested zones for reforestation or afforestation, to prevent landslides and hill degradation. Any expedition visiting Nanda Devi National Park should strictly observe the following don'ts, under observation of one forester guard, who should accompany each party into the Inner Sanctuary/Park to enforce ecological norms.

(1) The only approach path should be strictly guarded and no weapons allowed inside the Sanctuary, except hunting knives for self defence.

(2) No fire-arms should be allowed beyond Lata Village; all expeditions should deposit guns, pistols, revolvers etc. at Lata Post and retrieve them on return.

(3) Poachers, if any pass surreptitiously or by collusion, should be heavily penalised.

(4) No trekking party (Nanda Peak climbers apart) be allowed to remain inside the Inner Park for more than two weeks.

(5) The baggage (including sleeping bags and folded tents) should be searched for poached wildlife remains.

(6) No wood-fuel be used for cooking and visitors must necessarily carry Kerosene and stoves to cook.

(7) Only a small number of expeditions be allowed every year, which should consist of a limited number of persons.

(8) Special lectures of Mountain Ecology and Wildlife Conservation be held for the trekkers and mountaineers visiting Nanda Devi Sanctuary, either at Nainital Mountaineering Club (N.T.M.C) or at Garhwal University Srinagar en route.

Like a Bioreserve, the Nanda Devi National Park should consist of 3 areas : (a) a Core Area in the Inner Sanctuary proper; (b) a Buffer Area around the Core, consisting mainly of steep hill-sides fit for rock-climbing and mountaineering; and (c) a Cultural Area around the Outer Sanctuary zone. Scientific or Ecology students expeditions may be allowed only up to the Buffer area; while only the Nanda Devi climbers and ecology experts be allowed into the Inner or Core Area. This will obviate un-necessary removal or destruction of rare plants and animals from the old Inner Sanctuary.

N.T.M.C. (Nainital), famed for its rock and rope-climbing school, is arranging trekking parties of some college and University students from all over India this summer, as well as those belonging to mountaineering Clubs/Societies/Bodies. This is a magnificent opportunity for Indian youth to delve in mountain adventure and enjoy the stupendous landscape of Himalaya, learn to observe and conserve nature in its fascinating beauty and not pollute or destroy it. Our future generations will thank the present generation for maintaining the scenic splendour, inspirational, educational, cultural values and rare wildlife that Nanda Devi spells today.

The I.U.C.N. has defined a National Park as : " A large area where ecosystems are not altered significantly by man (such as Nanda Devi); and where geomorphology, fauna, flora and habitats are of great interest and beauty, which is protected by the highest *Central Competent authority* of the country; and where public is admitted for inspirational educational, cultural and recreational purposes". Thus it is a must that Nanda Devi National Park be protected by the forces of the Central Govt. of India, and not the State or Province where it lies.

REFERENCES

Anon. (1969) :*Conservation of Nature and Wildlife in India.* Govt. of India, Ministry of Agriculture.

Anon. (1969) : World conservation Strategy.

Anon. (1980) : *Report of the Committee for recommending Legislative Measures and Administrative Machinery for Ensuring Environmental Protection.* Deptt. of Science & Technology, Govt. of India.

Majupuria, T.C. (1986) :*Wildlife Wealth of India.* Tecpress, Bangkok.

Saharia, V.B. (1982) :*Wildlife in India.* Natraj Publisher, Dehra Dun.

Seshadri, B. (1969) : *Twilight of India's Wildlife.*

Rec. Adv. Fish Ecol. Limn. Eco-conserv.
III : 125-128

15

Biomass : Problems And Remedies

S. M. DAS

Sometime back it was pointed out by Dr. T.N. Khoshoo, Secretary, Department of Environment, Government of India, that a decentralised biomass energy system is most relevant to the rural environment in India; and that apart from its renewable fuel, food and fodder value it would also help in restoring the ecological balance in the country by improving water retention in the soil, improve the soil and maintain the carbon-dioxide balance in the atmosphere, which is increasing day-by-day.

Researches on regional stoppage of deforestation and methodology of afforestation and reforestation with regard to biomass, have only been started in the country during the last decade or so. But there is hardly any impact on this valuable renewable biological resource, although some relevant data are available today. The main drawback is that only sparse data are available at the regional, village and block levels. The forest departments, the Universities and the block development officers should all collaborate and plan out survey and research programmes to estimate the present biomass of renewable resource and the requirements and targets for each plan period in the country.

Next, the total plant productivity (dry organic matter per unit area and time) should be estimated as at the present time and after ten years of reforestation and afforestation of all ecologically tenable vacant barren lands. This will be determined mainly by the photosynthetic efficiency based on data of solar radiation (radiant energy) falling on the region concerned. The total biomass at any one time can then be calculated by using conversion factors described by Loomis and Williams (1963) which applies to our wildlife plant world. For biomass of animal wild life (including fish), on which we now depend mostly for protein food, the method of Sommerlate (1969) may be used, as the production of plant and animal life are interdependent ecologically both on land and in water.

The total annual biological yield (biomass, based on radiant energy as g/ha/year) are given in table I both for hills and plains :

TABLE - I: ***Biomass in g/ha/year***

Latitude °N			*Altitude Metres*		
0	-	3046.29	0	-	2281.98
10	-	2861.23	300	-	2270.18
20	-	2633.47	1000	-	2234.58
30	-	2281.98	2000	-	2198.98
40	-	. 2121.01	3000	-	2163.38
			4000	-	2127.78

If we consider the production of biomass in the world, the present main areas of production are:

Table II

Elements (regions)	*Area (million Square Km)*	%
The world oceans	300 M.S.q..Km	43%
The tropical forests	10 M.Sq. Km	29%
The Temperate forests	25 M. Sq. Km	10%
Grass land & Meadows	42 M.Sq. Km	10%
Farm land	14 M.Sq. Km	8%
Fresh Waters	8 M.Sq. Km	5%

It will thus be observed that the temperate forests have the largest area. This is available in India only in the Himalayas (from Kashmir to Kohima) and other hilly regions. It is, therefore, incumbent on us to concentrate our researches on hill forests and wildlife, both of which are being fast decimated by Man. This would require a thorough survey of the barren areas of the hills (man-made or natural erosion) and then afforestion with soil binding trees (*Oak, Aesculus, Cedrus* fir, *Rhododendron* etc., not single tap-rooted Eucalyptus, pine etc.).The high slopes should also be stabilised (upto 2000m) by planting shrubs such as multifooted *Lantano, Berberis*, Rhingal bamboo etc; and hill grasses such as lemon grass (*Cymbopogon*), terai grass (*Arundiella, Heteropogon*) upto 2,000-3,000 m; and *Agrostis, Andropogon, Poa*, etc. higher up. The grasses would not only bind the top soil and prevent erosion, but would also increase rapid production of biomass for cattle feed and fodder, while the shrubs would bind deep soil.

It has been estimated that, except for biological resources (biomass), most other natural energy resources are non-renewable. Some have stated (Fedorov 1980) that oil reserves would last only until 2092 A.D. and coal upto 2150 A.D. We must, therefore, augment our precious renewable resources (both plants and animals) by adopting some urgent steps for biomass production : -

(1) Logging and cutting of forests in damaged hills should be stopped for 10 years;

(2) Reforestation, afforestation, extension of grassy slopes and meadows; sufficient tree-planting in each block should suffice for the villagers fuel and fodder problems ;

(3) Setting up of new wildlife sanctuaries , reserves and National parks in different regions of the Himalaya, for research and study of biomass potential of wild-animals and plants, with steps to preserve gene pools; and domestication of some wild animals;

(4) Research into regional increase of carbondioxide in the atmosphere, due to less plants (and thus less photosynthesis) to utilise the CO_2; also due to fuel burning;

(5) Conservation of lakes, rivers, streams and springs and management of water resources, especially in the hills, which would increase biomass production of reeds, rushes, algae, aquatic weeds, fishes etc.

The wild animal biomess supply for protein food (including fish) should be thoroughly surveyed and levels of exploitation determined to conserve the animals. Rapid production of fish and establishment of wildlife forms (domestication) would augment valuable protein food for man. An estimate of animal wildlife biomass of areas in six different biotopes (Sommerlate 1969) is given below : -

Biotope	*Animal species*	*Biomass Kg/Km²*
Temperate mountain	Red deer (*Cervus elaphus*)	566-870
Forest (Upper Bavaria)	Red deer	
High and Moor (Scotland)	(*Cervus elaphus*)	1,000
Kashmir Dachigam	Hangul deer (*Cervus elaphus hanglu*)	250
Tropical Montane forest (e.g. Uttarkhand)	50 large and small	12,250/ Now less than 5000/
Savannah (Sarengeri)	22 large and small	5,000
Tundra (N.Canada)	Caribou	500

But by establishing preserves, reserves, sanctuaries, National parks, as also by production in enclosed wildlife farms and zoos, the biomass potential of animal biomass can be greatly increased. This has been done in some regions of Africa, U.S.A., U.S.S.R., and Germany.

The following are some mountain wild animals that may be raised in enclosed farms or domesticated for biological resources for man : Himalayan mouse hare (meat and fur), Hispid hare (meat and fur), Pigmy hog (meat, hide), Musk deer (Musk pod of commerce, each worth Rs. 4,000/ or more), Brown Goral (Meat hide, horn, milk), Tahr or Jharal (Meat, hide, horn, milk), Thar or serow (Meat, hide), Himalayan Ibex (Pashmina wool for finest shawls), Bharal or blue sheep (wool, fur, meat and horn), and the great Tibetan sheep (wool, horn). The biomass of these high-protein hill animals is higher than for members of the same family in the plains of India.

The proposal for biosphere reserves in India, which was mooted in 1970, has hardly been implemented in the hills. It would require intensive survey and research to locate the proper ecotopes for ecological conservation in biosphere reserves in the Himalayan ecosystem. The carrying capacity of proposed biosphere reserves has not yet been computed; and unless this is done, it may cause ecological imbalance, as has happened with the elephant in India and Africa.

The importance of survey and research into biomass in the mountainous area lies in the fact that nature sets an upper hand to fast growth up to timberline, which provides for our rainfall, water requirements (sub-soil water, springs and streams), for fuel and fodder in the hills foot hills. Along with a higher production of fresh shrubs, herbs and grasses, our forests can provide a biomass sufficient to meet the material requirements of Man. There are no biospheric capacity barriers to meet the food and energy requirements of man for the next 30 years, if ecologically exploited (Loomis and Williams 1963).

It is a good sign that two ecological battalions are being raised for the first time in the country for safeguarding the ecology of the Himalayan region, as announced by the Government. The two battalions were to be raised in Himachal Pradesh and Uttar Pradesh Himalayas. These should be composed of army jawans, University students (Part time), and PAC, BSF jawans(part time), along with forest guards, forest personnel and rangers.

REFERENCES

Asibay, EOA (1974) : Wildlife as a source of protein in Africa South of the Sahara. In Biological Conservation, *6* : 32-39.

Fedorov, E (1980) :*'Man and Nature'*, Progressive Publishers, Moscow.

King, J.M. & B.R.Heath (1975) : Game domestication for animal production in Africa *World Animal Rev., 16* : 23-50.

Loomis, J&R Williams (1963) : In :Richter. *Wild life utilisation and land use.*

Mossman, S.L.& AS Mossman (1976) : Wildlife utilisation & game ranching. *IUCN occasional papers, 17,* Morges, Switzerland.

Richter, W.von (1976) : Wildlife utilization and integrated land use in Botswana. *BOT 72/020. Field Document No. 4.* Botswana. Dept. of Wildlife Nat. Parks and Tourism.

Sommerlate , B. (1969) : In : Richter: '*Wildlife utilisation and land use.*'

Rec. Adv. Fish. Ecol. Limn. Eco-conserv.
III. 129-131

16

Is Nainital Lake Dying - Problems And Remedies

S.M.DAS

Ecosystem imbalance has now rapidly approached a point of no return in Nainital town and its environs. The once large placid beautiful Nainital lake is polluted to an extent that even good fish cannot live in it; while the coliform bacteria-laden water is hardly curable for potable purposes. The population has increased to an extent that housing and land falls far short of the human load and his metabolic and activity fallout in and around the town. More builidings would be a hazard to man himself in the town, as the peak load on the hillsides has been reached and, if loaded further, would cause landslides the like of which the town has not seen since the great landslides of the eighteen eighties, when the flats (playgrounds) were formed by a whole hillside killing over 120 Englishmen, women and children. This was primarily caused by deforestation and housebuilding without a scientific/ geological base. Within the year 1981 there have been scores of hillslides, roadslides and a major landslide on the route to Nainital, near Waldiya Khan.

The problem has become more acute due to the lakh or so summer visitors to the town and the consequent increase in effluents, detergents and sewage, much of which reaches the lake. Besides, increasing population has caused building of new houses without sanitary flush systems, relying on the old bucket/pot system, most of the excreta being hand-collected and thrown into the 24 or so steep nullahs surrounding the lake hillsides, resulting in the rapidly increasing pollution of the lake. The residual building material also caused high silting and consequent shallowing of the lake by about 6 meters in the last 50 years or so. A proposal was made to build tenements in the town for the increasing staff of the government and other offices being shifted anew to Nainital. There being little new safe land for these buildings, the old large bungalows with compounds were surveyed and promises made to purchase the open lands. The consequences can be easily imagined, when the flower beds and gardens would be totally destroyed and the fallout of refuse, litter, nightsoil vamp the hillsides leading finally into the lake at the first rains. The carrying capacity of Nainital town is now overloaded.

Four sittings of the Lakes Expert Committee and Hillside Safety Committee have been already held in the last two years, with the Commissioner of Kumaun as the Convener. Many suggestions for improvement by the D.S.T. (Govt. of India) expert (the present author) have been placed repeatedly before the Committee and incorporated in the reports. Some money has also been granted for laying new wide sewage pipes, roads and hillside improvement, as well as connecting all outfalls, sewage and effluents from high habitations to these sewage pipes. Urgent drastic action is required, based on the findings of the expert committee. If the grants fall short, more financial help may be secured from the Hill Development Board (U.P.), D.S.T. (Govt. of India) and Department of Environment (Govt. of India), since almost

52 crores have been ear-marked for hill development by the Planning Commission in the sixth plan outlay.

The present author had the opportunity to study the lakes of Kumaun and their environs in the project "Ecology of Kumaun Lakes and Development of Applied Environmental Research", granted to him by the D.S.T., for three years. The following are some of the recommendations after intensive study of Physiography, Climatology, Limnology and Biology of the lakes and environs (including catchment areas) of lakes Nainital, Bhimtal, Naukuchiatal, Sattal and Khurpatal.:

(1) The problem of rapid deterioration of Nainital town by overcrowding can be solved by creating a "Greater Nainital" on the past experience of expanding towns/ cities, as of "Greater London". New houses, buildings and tenements should now be constructed in sparsely populated Bhimtal, Naukuchiatal, Sattal and Khurpatal regions, which are connected by metal roads only 22 Km., 24 Km., 26 Km and 8Km. respectively from Nainital town, with regular bus service. Even some Govt. and private offices could be shifted there. The ideal surroundings of these places would also provide better life and living conditions for the staff. Further construction of buildings should be banned in Nainital.

(2) This,"Greater Nainital" can also be better planned for all amenities (Development and Management); and the tourist rush diverted to these four regions for better landscape, better water supply, clean air and spots for picnics and adequate aquatic sports and fishing (Lake Nainital has little sport fish today such as Mahseer, Hill-trout, Mirror Carp, etc.). Gentle slopes and vegetation in these areas would also obviate hazardous landslides, roadslides etc., if buildings are constructed on geologically sound lands.

(3) Plantation work be started in Nainital in all bare spaces with soil-binding trees and shrubs, such as *Quercus, Aesculus, Fraxinus, Vaughinia, Platanus*, Ringed bamboo, Willow, Berbery, and even *Lantana*, with spreading roots, and not *Eucalyptus*, Pine, Cedar, Cypress and Poplar which have non-spreading roots. Tree felling should be banned for at least 10 years. Mild slopes can be used for horticulture.

(4) The high slopes, including that of China peak (which had a massive landslide during the winter of 1980), should be buttressed specially at their toes and the outfalls. Lemon grass and other hill grasses may be planted on the steep slopes.

(5) Good quality water supply to Nainital town can be ensured by storing supply water in a large open reservoir, where colloidal impurities and bacteria can be coagulated and settled by alum and chemicals precipitated by sodium-metaphosphate. The high count of *E.coli* can be rendered harmless by using (a) bleaching powder, (b) potassium permanganate, and (c) exposure to air and sunlight which kills off *E. coli* in 2-3 weeks, as it is an inhabitant of the intestine of Man and some animals.

(6) Further silting of Lake Nainital can be prevented by constructing a silt settling tank at each nullah-mouth, in order to settle/deposit the silt and solid organic matter that may be removed once a week, especially during the rains.

(7) The old narrow-gauge (20 Cm) sewage pipes should be urgently replaced by wide gauge ones (60 Cm) and connected with all new-laid sewage pipes. The sewage should be carried out to an outfall outside the Nainital Valley. Sewage treatment facility should be provided at the outfall and the sullage used as a cheap fertilizer for vegetables and cash crops. Conservation should be strictly forbidden by law to throw refuse, nightsoil into the nullahs. A flush system should be made compulsory for all houses/tenements existing in Nainital town.

(8) The shallowing of the lake by 6 metres in the last 50 years (established bathymetrically in the D.S.T. Project)will cause the lake to be converted into a marsh in the next 100-150 years. Desilting operations are, therefore, a must and should be undertaken every five years or so.

(9) There is a 6-inch layer of organic putrifying mat over the lake bottom, which causes anaerobic condition at lake bottom with evolution of ammonia, hydrogen sulphide and methane, that increase and reach the surface waters every winter (dissolved Oxygen falling to 2.5 mg/1.), causing mass mortality (80,000 dead fishes observed) of lake Nainital fishes. Unless this layer is reduced by dredging (at least 4 inches) and further organic pollution of the lake stopped (by using thick wire-tilting net traps devised by the present author to sieve off leaf-litter, debris, rubble and detritus and tilted into trucks of carriers to be disposed outside valley), the organic pollution in the lake will go on increasing. A further control can be imposed by spreading slaked lime (one quintal/hectare) over the lake surface every autumn.

(10) Tourist trade can be enhanced by extending Nainital to "Greater Nainital" to benefit the common man all over the district. Recreational facilities (boating, angling,water-skating, swimming-diving etc.) should be provided. But shooting should be strictly prohibited, as little wildlife is left in these Kumaun hills today. Planting of ornamental and flowering plants around the lake-shore would not only relieve the monotony of the Tallital - Mallital Mall road walk, but also awaken the aesthetic faculty of Nainital residents. This beautification should also be done around Bhimtal, Naukuchiatal, Sattal and Khurpatal in "Greater Nainital ".

We have had the bounty and beauty of Nainital for over a century. Let not our descendents say we destroyed it, for surely Nainital as a beauty spot in Kumaun Himalayas is dying a slow but sure death.

REFERENCES

Das, S.M. 1969. Some aspects of fish ecology and limnology in Western Himalayas. UNESCO. *Rep. Seminar on Ecology of Tropical Highlands, Kathmandu, Nepal* : 15-27

Das, S.M. 1978. Development versus Environment in the Himalayan region. The Himalayas. *Garhwal University Journal* : 39-43.

Das, S.M. 1979. Effects of increased landuse upon freshwaters in the Central Himalayas. *Progressive Water Technology*, 11(6): 195-208.

Das, S.M. 1981.Final report on ecology and fisheries of Kumaun lakes. *Report D.S.T.*, Govt. of India (1981) : 1 - 108.

Das, S.M. and L.S. Khanka 1979. Physiography and hydrography of Nainital lake. *Geographical Review, India*, 41 (2) : 139-143.

Das, S.M. and Jyotsna Pande, 1978. Some Physico-chemical and biological indicators of pollution in lake Nainital, India. *Indian Journal of Ecology*, 5(1), 7-16.

Das, S.M. and Jyotsna Pande, 1982. Pollution, fish mortality and environmental parameters in Lake Nainital. *Journal of the Bombay Natural History Society*, 79(1) : 100-109.